New
Dimensions
in
Military
History

New Dimensions in Military History

An Anthology

Edited by

Russell F. Weigley

PRESIDIO PRESS
SAN RAFAEL · CALIFORNIA

NEW DIMENSIONS IN MILITARY HISTORY
An Anthology Edited by Russell F. Weigley

Copyright © 1975
PRESIDIO PRESS
1114 Irwin Street
San Rafael, California 94901

All rights reserved. No part of this publication may be reproduced, stored in a retrieval system, or transmitted in any form or by any means, electrical, photocopying, recording, or otherwise without the prior written approval of the publisher.

Library of Congress Number
76-4159
ISBN 0-89141-002-3

Book Design by Wolfgang Lederer
Printed in the United States of America
by Mowbray Company, Providence, Rhode Island.

For all those who
'save what remains
not by vaults and locks,
which fence them from
the public eye and use
in consigning them
to the waste of time,
but by such a
multiplication of copies
as shall place them beyond
the reach of accident.'

THOMAS JEFFERSON

CONTENTS

Introduction
RUSSELL F. WEIGLEY
Temple University
page 1

I The Nature of Military History
Military History: An Academic Historian's Point of View
JAY LUVAAS
Allegheny College
page 18

II Armed Forces and Society
Armed Forces and Society: Some Hypotheses
THEODORE ROPP
Duke University
page 38

Cultural Change, Technological Development and the Conduct of War in the Seventeenth Century
DANIEL R. BEAVER
University of Cincinnati
page 73

III National Security and Military Command

French National Security Policies, 1871–1939
RICHARD D. CHALLENER
Princeton University
page 92

The Japanese Army Experience
ALVIN D. COOX
San Diego State University
page 123

High Command Problems in the American Civil War
WARREN W. HASSLER, JR.
Pennsylvania State University
page 152

American Command and Commanders in World War I
EDWARD M. COFFMAN
University of Wisconsin
page 176

IV The Composition of Armed Forces

Conscription and Voluntarism: The Canadian Experience
R.H. ROY
University of Victoria, British Columbia
page 200

The Multi-cultural
and Multi-national Problems
of Armed Forces

RICHARD A. PRESTON
Duke University

page 225

The Army of Austria-Hungary,
1868–1918:
A Case Study of a Multi-ethnic Force

GUNTHER E. ROTHENBERG
Purdue University

page 242

V Armed Forces in Politics and Diplomacy

The Petrograd Garrison and
the February Revolution of 1917

WARREN B. WALSH
Syracuse University

page 256

Multilateral Intervention in Russia,
1918–1919:
Three Levels of Complexity

GADDIS SMITH
Yale University

page 274

Military Government: Two Approaches,
Russian and American

EARL F. ZIEMKE
University of Georgia

page 290

VI Unpopular and Unconventional Wars

The War of 1812 and the Mexican War
HARRY L. COLES
Ohio State University
page 311

A Case Study in Counterinsurgency: Kitchener and the Boers
THOMAS E. GRIESS
United States Military Academy
page 327

Revolts Against the Crown: The British Response to Imperial Insurgency
J. BOWYER BELL
Columbia University
page 358

VII Some Conclusions About Military History

The Nature and Scope of Military History
MAURICE MATLOFF
U.S. Army Center of Military History
page 386

Military Leadership and the Need for Historical Awareness
HAROLD K. JOHNSON
General, U.S. Army, Retired
page 412

New
Dimensions
in
Military
History

Introduction

RUSSELL F. WEIGLEY

THE GREATEST of the schools of professional military higher education have always given military history a central place in their curricula. Most conspicuously, the German *Kriegsakademie*, the first and most famous of all war colleges, in its heyday under the guiding influence of the elder Helmuth von Moltke based its study of strategy upon military history. Moltke believed there could be no other foundation for the education of a strategist. Under Moltke's influence, furthermore, the Prussian and German General Staffs meticulously compiled historical documents and studies of all their army's campaigns for the guidance of commanders in future campaigns. Thus steeped in military history, the German high command went on to lead its forces to disastrous defeat in two world wars.

So it is appropriate to be modest about the utility of military history. Because the opportunity for the soldier to practice his profession must necessarily and fortunately be limited, military history is an indispensable substitute for direct experience in preparing the soldier for the crises of leadership. But it is a highly imperfect substitute. The vicarious experience of war in the security of a study or an armchair so little approximates the pressures of combat that it is hardly experience at all. Such is the first and most obvious difficulty, but there are other, more subtle ones.

The professional soldier seeking in history preparation to meet his responsibilities, of administration as well as of strategy and tactics, naturally expects that history should provide him with practical principles to guide action—with lessons. Other-

INTRODUCTION

wise, why should a busy man bother with it? But the professional dispenser of history, the historian, habitually shies away from a promise to teach lessons. The historian's own professional education stresses the ambiguity both of the evidence from which he must reconstruct the past and of the conclusions to be drawn from the evidence. His conditioning prepares him to find in the past not lessons to guide conduct, but only the most ambiguous, the messiest, of meanings. Military history *was* the foundation of the professional education of the German high command; military history *did* help cultivate in the German command the breadth and depth of strategic perception that contributed indispensably to Germany's holding the whole world at bay for four years and then again for six years in two world wars; but it was their very sense and understanding of history—particularly their awareness that they, the military, had always been the core of the Prussian state and that Prussia's greatness had always rested on arms—that tempted the German military leaders to try to grasp too much by military means, and brought nemesis and disaster. History can misguide as readily as it can guide.

Because the historian's response to the military's quest for lessons in history has tended to be unsatisfactory, soldiers have often looked elsewhere, to seek nonhistorical aids to their professional education. Moltke and the German military educational system were in the tradition of military studies that stemmed from Karl von Clausewitz and that sought to understand war by probing its moral and psychological dimensions with history as a basis. Claùsewitz himself, more the philosopher than the practical soldier—though he had ample practical experience of Napoleonic war—accepted and in fact reveled in the ambiguities uncovered by a historical survey of war in search of war's moral essence. His masterwork *Vom Kriege* is a kaleidoscope of ever-shifting and ever-ambiguous observations—which makes it forever a source of frustration and fury to the practical soldier who decides to go beyond giving it lipservice and actually to read it. Thus, however often Clausewitz's philosophical

INTRODUCTION

subtlety is praised, soldiers (except possibly in Clausewitz's own country) have consistently recognized the superior utility in Clausewitz's contemporary and principal rival as an interpreter of Napoleon and founder of modern military thought, Antoine Henri, Baron de Jomini.

For Jomini offers generalized rules that the strategist, logistician, tactician, and even the military administrator can follow in practice, handy injunctions regarding what he should attempt to do and what not to do. With Clausewitz, the reader is constantly trying to disentangle the everyday realities of war from the absolute essence, and emerging with no firm, definable conception of either. With Jomini, the reader learns self-evidently applicable principles—that the object of strategy, for example, is "To throw by strategic movements the mass of an army, successively, upon the decisive points of a theater of war, and also upon the communications of the enemy as much as possible without compromising one's own."[1] Here is a statement that in fact gives the commander a notion of what he is to do, and a point of departure from which to turn from study to activity. Jomini points straight toward the twentieth-century codifications of the principles of war. But while Jomini, like Clausewitz, began with a study of military history and particularly of the campaigns of Napoleon (in fact, in far more specific detail than any historical instances in Clausewitz), Jomini achieved his didactic straightforwardness and utility by abandoning historical ambiguity for generalizations derived from geometric conceptions of the theater of war and the battlefield. It is the Clausewitzian tradition of military education with its muddy philosophical waters that is essentially historical in its approach. The Jominian tradition, which reached America through Dennis Hart Mahan at West Point and proceeded through Captain John Bigelow's *Principles of Strategy* of the 1890s to a current work such as Rear Admiral Henry E. Eccles' *Military Concepts and Philosophy*, is less a historical tradition of military education than a geometric and in a larger sense a scientific one; it seeks the Newtonian natural laws of command.[2]

INTRODUCTION

And if a Newtonian approach to a science of war may seem antique in its conceptions of what is scientific, the years following World War II have also offered a far more modern scientific alternative to history as a foundation for the study of war. With the advent of the nuclear age and nuclear strategy, "hard" scientists and mathematicians turned to the study of military problems. Some of them did so at first out of a sense of special responsibility for having dramatically worsened the potential horrors of war; *The Bulletin of the Atomic Scientists* expressed this motivation and attempted as a means of meeting responsibility—or of assuaging guilt—to apply the methods and logic of science and mathematics beneficially to public and military policy. Subsequently, the continued rise of both nuclear and missile technology made military problems seem increasingly matters for scientific research, tying the strategy of nuclear deterrence and nuclear balance ever more closely to physics and mathematics. At the same time, the growth also of computer technology further encouraged a mathematical approach to strategy, suggesting efforts to resolve strategic problems into segments that could be expressed quantitatively and thus could be subjected to computer analysis. Allied to the growth of computers was the development of operations research and operations analysis and their evolution into systems analysis as approaches to the interlocked issues of strategy, tactics, and weapons systems. Operations research began in World War II as the application of systematic and especially quantitative analysis to the problem of securing optimal performance from weapons. By the end of the war, operations research had led into operations analysis, a term indicating a larger study in which the quantitative methods of operations research would be only one among all appropriate methods of systematic investigation of problems in military technology—but with the emphasis, especially in the American version of operations analysis, still on quantification. Finally, an expansion of operations analysis to analyse the uses of future as well as existing weapons systems, and in the process to rationalize the choices among various

INTRODUCTION

strategies that are implicit in choices among weapons systems, became systems analysis. All these developments made nuclear age military study in unprecedented measure the province of the mathematician, of the physicist, or at least of the quantifying social scientist.

Soldiers of the past could hardly be blamed for preferring Jomini's geometric and generalizing approach to military study over Clausewitz's historical and philosophical ambiguities; Jomini and his followers could give them more immediate help than Clausewitz and his followers. It is not surprising either that military men after World War II should have hoped for more help from the new scientific and mathematical strategists than from the historians. The ambiguity of historians' verdicts was discouraging enough; now some historians lent futher discouragement by disparaging the relevance of their own discipline in the nuclear age. When writing what is still the best survey of American military history, *Arms and Men*, Walter Millis concluded in 1956 that "The advent of the nuclear arsenals has at least seemed to render most of the military history of the Second [World] War as outdated and inapplicable as the history of the war with Mexico."[3] The expression of such views by other historians was not uncommon. The propensity of historians to take short views when they of all people should be taking a long view is indeed another cause for modesty about the uses of history; no less a personage than Thuycidides commenced this unfortunate tradition among military historians when he asserted that nothing that had occurred in Greece before his own time was worth recording.[4]

Still more unfortunately, however, the hard sciences and mathematics proved to offer at least no better guide to military policy and strategy than had military history. The evidence offered by the new scientific strategists might seem reassuringly firm in its mathematical precision, but their conclusions about a proper overall strategy for the nuclear age soon developed a disconcerting tendency to swing pendulum-like back and forth between "finite deterrence" and "infinite deterrence," or those

INTRODUCTION

concepts under other names. Still more disconcertingly, the systems analysts and their computers, when ensconced at the center of the Department of Defense by Secretary Robert S. McNamara, largely promised prompt and satisfactory results at acceptable cost for an American military intervention in Indochina. The unhappy Indochina experience is one in which, in retrospect, it would seem that history might have offered more useful guidance, for the American failures were rooted in a lack of historical sense of the differences between American culture and a very different, very history-bound culture.

By the time I spent the academic year 1973–1974 as a visiting professor at the U.S. Army Military History Research Collection at Carlisle Barracks, teaching and counseling students of the Army War College which is located at the same post, a desire to return to history for guidance in military policy and strategy was coming to be frequently expressed among faculty and administration and especially among students at the War College, with disillusionment over the once-inflated claims of quantitative approaches to policy and strategy being explicitly stated, in terms much like those of the preceding paragraph. By that time, too, the curriculum of the Naval War College at Newport had already for a year been standing again on the historical foundations that Rear Admirals Stephen B. Luce and Alfred Thayer Mahan had initially fashioned for the college, the senior course proceeding from considerations of national and military strategy through management to naval tactics, all—and especially strategy—treated from a historical perspective.

One important reflection of a renewed interest in military history as a foundation for the military man's professional education and professional development was the establishment in January, 1971, of the Department of the Army Ad Hoc Committee on the Army Need for the Study of Military History, whose report in four volumes was published the following May.[5] This committee noted the decline of the Army's and the other services' concern for military history after World War II, suggested some of the explanations for the decline in the new problems and

INTRODUCTION

the new investigative and conceptual tools of the postwar world, but reaffirmed the value of the study of history—and beyond that, the value of historical mindedness—for the military professional. The report called for fuller use by the Army of its already existing historical agencies, especially the Office of the Chief of Military History (now the Center of Military History), which had developed principally out of the project of writing an official army history of World War II but had ongoing functions as a keeper of historical records and an agency for writing army history, and the Military History Research Collection, founded in 1967 to be a single center for the holding of the Army's library materials of historical importance.

The report of the Ad Hoc Committee noted that, in something of a paradox, the study of military history in civilian universities had been growing during the very postwar years when the armed services' interest was declining. The insecurities of the postwar world had generated a new civilian interest in national security studies, largely among scientists and social scientists but in which the academic discipline of history had shared. The committee proposed that a rapport be nourished between the Army and the civilian military historians for the mutual advantage of both. In particular, it recommended the establishment of three Army chairs of military history at West Point, Carlisle Barracks, and the Command and General Staff College at Fort Leavenworth, which visiting civilian military historians would hold for one-year terms: the first two of these chairs were created in 1972–1973, the third in 1974–1975. Overall, the Ad Hoc Committee recommended an enlarged place for military history in the curricula of the service schools, and it encouraged a return to military history in the individual study of the professional soldier.

In the autumn of the same year that the Ad Hoc Committee deliberated and reported, 1971, historians of the Research Studies Division of the U.S. Army Military History Research Collection inaugurated, at the request of the Commandant of the Army War College, an elective course in military history for

INTRODUCTION

Army War College students. Such a course had already been contemplated before the Ad Hoc Committee reported, but it met the committee's recommendation that "Challenging elective courses in military history should be offered at all service schools."[6] Though the military history education of officers rising in the post-World War II Army had suffered a general neglect, an introductory course in military history did not seem appropriate to the Army War College, the Army's senior service school, where the students are senior lieutenant colonels and junior colonels averaging forty-two years in age and twenty years in service. Rather, the military history elective sought to acquaint students with the variety of the latest research findings and interpretations in military history by bringing to Carlisle Barracks for lectures and formal and informal discussion the leading American military historians, with the staff historians providing continuity. In a limited time within a crowded curriculum, the main purpose was to encourage the historical mindedness emphasized by the Ad Hoc Committee rather than to impart a specific body of knowledge—to acquaint the students with the dimensions and potential of military history—thus the title of the course and of this book. Hoping similarly to offer the reader a sense of the current condition of military history, to suggest its variety and its possibilities, this anthology comprises a selection from the presentations given during the first three years, 1971–1974, of the Military History Research Collection-Army War College New Dimensions course.

I myself lectured in the first two years of the course and as visiting professor was present for the whole of the third year's offering. My participation in this course affords one of the principal sources for my earlier statement that Army War College students have expressed a strong inclination to return to military history as a foundation for professional study. At the Army War College, this inclination is now being met by additional electives in military history and by injecting a growing historical quotient into all areas of the curriculum, especially those concerned with national and military strategy. But the students I have met both

INTRODUCTION

at Carlisle Barracks and on visits to other senior service schools have also expressed reservations, and describing the apparent reversal of the American armed forces' recent indifference to military history should not be an occasion for abandoning a suitable modesty about the uses of military history.

For one thing, as professional officers the students still frequently press historians to define the lessons to be learned from the study of military history. They still want something of practical, professional utility. In seeking it, they are still apt to receive from those who are professionals in history rather than in arms—from this historian among them, on several occasions—the frustrating reply that history does not teach lessons. But that reply will not do. If the meanings of history are invariably messy, so are the meanings of experience in our individual lives; but surely experience offers some guidance for conduct. If the "lessons" of history are more often than not misused, especially by those who lack historical mindedness and thus the sense of history that throws into perspective the mere facts of the past, nevertheless misuse does not imply that there are no lessons at all.

Jay Luvaas in the first of the essays printed here, while complaining of the military penchant to seek "precise, hard lessons which, once mastered, will lead inevitably to success on the battlefield," goes on to discuss how Sir Basil Liddell Hart did draw from history, especially from Genghis Khan and William Tecumseh Sherman, the lessons in the proper use of mobile columns that shaped his conception of what became the *Blitzkrieg* of armored columns. The lessons may be more likely to be negative ones, cautionary examples of what not to do. But there is surely value also in being saved from the repetition of mistakes. Knowing how the Italian campaign of World War II demonstrated the severe limitations of attempted aerial interdiction of the enemy's lines of communication, even against lines traversing mountains and leading into a narrow peninsula, and with the would-be interdictors enjoying almost complete command of the air, might have been expected to save General

INTRODUCTION

Douglas MacArthur's headquarters from inflated expectations of aerial interdiction in Korea; knowing how the Korean War again demonstrated the severe limitations of aerial interdiction under similarly favorable circumstances might have been expected to deflate excessive expectations thereafter. If the lessons are not always read, nevertheless they exist.

It is especially in offering insights for the guidance of strategy that military history can be expected to be useful. No doubt nineteenth-century military historians exaggerated the extent to which, while changing weaponry makes for changing tactics, the principles of strategy remain stable. Admiral Mahan, who especially emphasized the stability of the rules of strategy through all changes in tactics and technology, failed as a strategic prophet to the extent that he did not recognize the impact of the new technology of the submarine and the automobile torpedo upon strategy as well as tactics, in undercutting the strategic value of conventional command of the sea. On the other hand, the United States Navy was still able to conduct virtually a textbook demonstration of Mahan's strategic principles in applying them to defeat the maritime Empire of Japan in 1941–1945. If Clausewitzian ambiguities are of limited use to the would-be strategist, so, beyond a few basic principles, are Jominian geometries. Jomini's definition of the object of strategy has a self-evident utility for the practical soldier that Clausewitz's dialectics lack; yet it is also true, as Colonel F.N. Maude, a turn-of-the-century British military critic, said of Jomini's teachings, that:

> These ideas are, after all, elementary, and readily grasped even by the average intellect, though many volumes have been devoted to proving them, and yet they are all that Jomini and his followers have to offer us—a fact that both explains and justifies the contempt with which military study was so long regarded by practical soldiers in England.[7]

Between the horns of the dialectic of Clausewitz and the geometric oversimplicities of Jomini, military history remains the best guide to which the strategist can resort, the military man's

INTRODUCTION

only substitute for the experience through which the practitioners of other professions continually learn, but which is so necessarily denied to the military most of the time.

To emphasize a military history that can offer guidance to strategy is to emphasize a military history that deals with conflict. Perhaps the most conspicuous shortcoming of the lectures thus far offered in the New Dimensions in Military History elective at Carlisle Barracks, and thus of the essays in this anthology, is that too few of them deal with conflict. The military history of the civilian military historians who interested themselves in national security problems under the influence of World War II and the Cold War has been pridefully called by its apostles a "new military history." Previously, military history dealt almost exclusively with battles and campaigns. The "new military history" is new in its concern for military history as a part of the whole of history, not isolated from the rest, for the military as a projection of society at large, for the relationships of the soldier and the state, for military institutions and military thought. But the salutary enlargement of military history's domain when it encompasses such subjects becomes marred by pretentiousness—or preciousness—when the new military historians so completely assert their liberation from old-fashioned drum-and-trumpet military history that they fail to deal with it at all. Military institutions have no reason for being except to engage in conflict or the implicit conflict of deterrence. Ultimately, conflict is what military history is all about, and the tendency of military historians sometimes to pretend otherwise is another cause for modesty in discussing the whole endeavor.

If the military historian deals with conflict as he should, he will of course enter the realm of tactics as well as of strategy—the tactical arena where even Admiral Mahan, whose appreciation of the influence of technological change was limited, acknowledged that new technologies render everything unstable. Even here, however, the relevance of a past as remote as the Mexican War is not quite so simply dismissed as our quotation from Walter Millis might have implied. As we enter the era of

INTRODUCTION

precision guided weapons, the evolution of tactics may be making a quantum jump from the tactics of World War II. But such a transformation may render all the more worth studying the earlier quantum jump that occurred between the Mexican War and the American Civil War, in the tactical transition from the smoothbore musket to the rifle as the standard infantry arm. The increase in precision was proportionately similar to that now offered by precision guided weapons, and we may well look to the first nineteenth-century battlefields that felt the power of massed rifles to help form judgments of the probable impact tactically—and psychologically—of the new precision weapons of our own day.

Furthermore, the first cries that nuclear weapons had left previous military history a matter of merely antiquarian interest went up before the shape of war in the nuclear age had revealed itself. They rose before a nuclear balance of terror between rival powers made conflict utilizing nuclear weapons highly unlikely, but in doing so pushed military conflict back into simpler forms. Wars have occurred in the nuclear age, but the restrictions imposed by the balance of terror upon their scope and form have made earlier conflicts yet more primitive in weaponry than the Mexican War nevertheless relevant to an understanding of them—earlier conflicts such as the one in Spain that first bore the name "guerrilla," or the war of national liberation waged by the American partisans after British reconquest of Georgia and the Carolinas during the War of American Independence. And as nuclear deterrence has confined the actual invocation of military force to prenuclear forms, armed forces have retained in the nuclear age much the same organization, traditions, and values, and much the same kinds of relationships with the civilian societies around them, that they have had for several centuries past. Thus the "new military history's" investigations of the soldier and the state and of armed forces and societies continue to have relevance to current civil-military issues. The experiences of the United States with both professional and conscript armies in the past can still throw light on the probable outcomes

INTRODUCTION

of the new experiment with a professional, nonconscript army in the present.

The very imperfections of military history have much to do with the continued importance of studying it. If men could indeed readily grasp lessons from out of the military past, they would not so often entangle themselves in wars that turn out to be futile, if not disastrous, for their authors. (This is to say nothing of the yet much deeper roots of war in the imperfections of the human character.) But because the meanings of history are messy meanings, and states go on blundering into wars, every state must be prepared to employ force from time to time in its relations with other states, and soldiers must prepare themselves through the study of history and by other means to practice their profession. As this introduction is written in 1975 against the background of the Communist conquest of Indochina, Americans almost unanimously protest "Never again!" when they think of their recent Indochina War. But short of renouncing military force, how can we be sure of never embarking on such a venture again? Which international problem will be analogous enough to apply the messy meanings drawn from Indochina?

The utility of military force as an instrument for accomplishing national objectives has become severely diminished during the past two centuries. The threat of an utter nuclear holocaust tends to preclude the use in combat of any nuclear weapons whatever, lest their employment escalate into holocaust. But ever since the American Civil War, the firepower of "conventional" weapons has been so great that whenever resolute opponents of anywhere nearly equal strength within the contested theater of war meet on the battlefield, the outcome has tended to be a tactical stalemate that becomes a strategic stalemate as well. The prolonged tactical and strategic deadlock of the American Civil War reappeared in the Russo-Turkish War of 1877–1878, the Russo-Japanese War of 1904–1905, and most horribly in World War I; the airplane and the tank merely seemed to break the deadlock early in World War II, the dead-

INTRODUCTION

lock's real persistence coming to be demonstrated by the necessity for a prolonged war of attrition on the Eastern Front to resolve the war; and deadlock set in again in Korea and Indochina, and as the Yom Kippur War of 1973 suggests, was absent from the earlier Arab-Israeli wars mainly because of the disparity between the strength and resolve of the adversaries.

In these conditions, military force can no longer be an instrument for nicely calculable employment to a state's positive advantage, as when in the eighteenth century Frederick the Great could employ force for the surgical removal of Silesia from Austria. Military decision at small cost is no longer often available, except against the weakest adversaries and with diplomatic risks even then.

But if we can no longer expect to enjoy much positive gain from the use of military force, nevertheless we must be prepared to resort to force if only to prevent loss. The gains of the United States from the Korean War were of this negative sort only—the mere prevention of loss; but if we consider nothing else but the strategic position of Korea pointing toward Japan, such a result was no small thing. The necessity for retaining armed forces and the likelihood of having to use them are with us still; and so, then, are the uses of military history to illuminate the path both of the professional soldier and of the citizens who employ him.

Writers often regard editors as dispensable; this editor has been dispensable to more than the customary degree. Not only did the writers do the real work, they then graciously granted permission to print the papers that they had prepared originally as lectures to the U.S. Army Military History Research Collection-Army War College New Dimensions in Military History course. In addition, the book fails to reflect adequately the credit belonging to the initiators and organizers of the course. The design and organization of the New Dimensions course in the autumn of 1971, when it was first given, were the work of two able military historians of the staff of the Military History Research Collection, Drs. Don Rickey, Jr., and B. Franklin Cooling. After Dr. Rickey departed Carlisle Barracks, Dr. Cooling

INTRODUCTION

continued primarily alone to assemble the speakers and organize and administer the course for the two subsequent years that are also represented by papers in this anthology. The merit of this collection rests in the first place on Rickey's and Cooling's organization of the course, knowledgeable selection of lecturers, and capable management of the course.

Colonel George S. Pappas has played a major role in creating the anthology at both the inception of the papers and their appearance in print. He was the first director of the U.S. Army Military History Research Collection under whose auspices the New Dimensions course was given—and much more than Director, the founder of the Collection who transformed it from an idea into one of the leading military history research centers of the world. Now retired from the United States Army, as President of the Presidio Press he proposed the anthology and encouraged it through every stage of publication. He saved the editor the larger part of the drudgery, especially of corresponding back and forth with the contributors, that might have accompanied his work.

Colonel James Barron Agnew, the present Director of the U.S. Army Military History Research Collection, both granted the permission of the Collection for the publication of this anthology and ensured that there were no Army restraints or objections that might otherwise have been overlooked. Colonel Agnew's agency has already published for limited circulation under its own imprint all the lectures given during the first three years of the New Dimensions course; by encouraging the publication of this anthology, Colonel Agnew has generously fostered a wider distribution of representative lectures. Brigadier General James L. Collins, Jr., Chief of Military History, merits thanks for similar helpfulness.

To all those named above the editor is grateful; the anthology required the contribution of every one of them.

R. F. W.

Philadelphia
June, 1975

Notes

1. Baron de Jomini, *The Art of War*, tr. G.H. Mendell and W.P. Craighill (Philadelphia: Lippincott, 1862; reprinted Westport, Conn.: Greenwood, 1971), p. 70.
2. John Bigelow, *Principles of Strategy, Illustrated Mainly from American Campaigns* second ed., revised and enlarged, (Philadelphia: Lippincott, 1894; reprinted New York: Greenwood, 1968); Henry E. Eccles, *Military Concepts and Philosophy* (New Brunswick: Rutgers University Press, 1965).
3. Walter Millis, *Arms and Men: A Study in American Military History* (New York: Putnam, 1956), p. 7.
4. Thucydides, *The Peloponnesian War*, trans. Thomas Hobbs, ed. David Grene (Ann Arbor: University of Michigan Press, 1959), bk. i: 1–2.
5. *Department of the Army Ad Hoc Committee Report on the Army Need for the Study of Military History* 4 vols. (West Point, 1971).
6. *Ibid.*, I, 50.
7. *Encyclopaedia Britannica* eleventh ed., 29 vols. (New York: Encyclopaedia Britannica Co., 1910–1911), XXV, 992.

I
The Nature of Military History

Military History:
An Academic Historian's
Point of View

JAY LUVAAS

page 18

═══════════ COMMENTS ═══════════

An introduction to military history for senior officers in whose post-World War II military education this discipline had been lacking was the central purpose of the New Dimensions course. Customarily, the course began with a visiting historian who surveyed the whole field. Jay Luvaas of Allegheny College had such a survey as his assignment when he was the initial visitor in the second year of the course, in the autumn of 1972. His judgments about the values of military history differ enough from what the editor has just said to be worthy of notice simply as another point of departure for the essays to follow; but Professor Luvaas' stature in the field offers a much more important reason for noticing his remarks about the nature of military history.

Military history must be international history; the military historian cannot limit himself within the boundaries of a single nation-state as historians in other specialties are too much wont to do. Jay Luvaas exemplifies the international nature of the military historian's craft by combining special expertise in both modern British military history and the American Civil War. His books include The Education of an Army: British Military Thought, 1815–1940 *(Chicago: University of Chicago Press, 1964) and his edition, with much astute commentary of his own, of some of Colonel G. F. R. Henderson's studies in military history,* The Civil War: A Soldier's View *(Chicago: University of Chicago Press, 1958). Professor Luvaas was the first occupant, in 1972–1973, of the visiting professorship in military history at West Point.*

Military History:
An Academic Historian's Point of View

JAY LUVAAS

MILITARY history, to echo Alexander Pope's description of women, is "at best a contradiction still." It is at once the oldest form of history and the most recent to gain academic respectability. Indeed, in a very real sense history *began* as military history, for the frequent wars in classical times provided a popular theme for the historian no less than the poet. Herodotus, the "father of history," gave Greek warfare an epic quality in his account of the Persian Wars; Thucydides was a military historian of the first rank; and one has only to think of Xenophon's *Anabasis*, Caesar's *Commentaries*, and vast portions of Polybius, Plutarch, and Livy to appreciate the significance of military history in the ancient world.

Yet only in our own time has military history been permitted a place in the ranks of departmental offerings in most colleges and universities. Professor R.M. Johnston taught a "half course" in the subject at Harvard in 1912—which was the only course of its kind in an American university at the time. There were still just two universities twenty-three years ago that offered graduate work in the subject. Yet today by actual count there are 109 institutions of higher learning offering specialized courses in military history, not counting the offerings in any ROTC department. While prestigious universities of the caliber of Dartmouth, Duke, Harvard, Cal Tech and MIT, Michigan, Princeton, Rice, and Stanford are active in the field, most of the schools offering courses in military history are recently estab-

THE NATURE OF MILITARY HISTORY

lished or expanded state universities and colleges. Liberal arts and technical institutions have lagged behind, and more often than not the department that does offer a course in military history is influenced more by a particular professor's special interest than by a belief in the value of the subject.

Another bewildering contradiction: as the published report on the Army need for the study of military history has recently stated, "while civilian interest in military history has increased, the Army has shown less interest in teaching the subject in service schools than it did before World War II!" This becomes all the more curious when it is realized that while it has required a concerted and organized effort on the part of the American Military Institute to persuade program committees for the various professional historical associations to allow sessions on military history a regular part of the annual program, Army leaders have come increasingly to view the subject in a truly liberal arts context. Disdaining the so-called "drum and trumpet" school that replayed the campaigns of the last century with all the zeal and finesse of a John Phillip Sousa march, military establishments today are concerned not only about operational topics, but more especially about history dealing with the administrative and technical aspects of war and the relationship between the military, social, political, economic, and psychological elements at the national level. Thus we have the interesting paradox of many academic historians looking down their scornful noses at the colleague with an interest in military history instead of diplomatic history or race relations, while the military profession is not only broadening its view of the subject but reaching out to involve professional historians in every conceivable way. Academicians, who are not always the most tolerant of individuals, retain a lingering suspicion that military history is somehow to be identified with "militarism," and this at a time when the Army is officially recognizing the need "to develop historical mindedness among the officer corps at large and to contribute individually to broadened perspective, sharpened judgment, increased perceptivity, and professional expertise."

AN ACADEMIC HISTORIAN'S POINT OF VIEW

Military history has always fallen between two schools. As military practice expanded and became more specialized during the Renaissance, so did the literature on war. Machiavelli, by opening a new line of inquiry into the nature and functions of the state and the relations between warfare and politics, was the first to view "soldiering as a branch of citizenship and warfare as a branch of politics."

But as the years progressed and the wars intensified, military history became the exclusive province of soldiers. The great military writers of the eighteenth century drew little distinction between military history and military theory—Marshal de Saxe in his *Reveries*, Guibert in his *Essay upon Tactics* and William Lloyd in his *History of the Late War in Germany* used history as a base for their theories. Jomini, the greatest military theorist of the post-Napoleonic period, insisted that the true function of military history was "to develop the relations of events with these principles," a judgment that was accepted by former soldiers who monopolized the field of military history. This was true of Sir William Napier, author of the classic *History of the War in the Peninsula*. It is seen more clearly still in the books of a retired British general named John Mitchell, who took a particularly pragmatic view of the subject. Military history, Mitchell wrote twenty years after Waterloo, should furnish practical lessons; otherwise it was "useless and unprofitable." To focus attention on the drama of battle or the details of a campaign at the expense of other determining factors would probably lead to a superficial view of what actually had happened. And thus to misunderstand the lessons of history would give rise "to new and fashionable doctrines, that are followed and upheld till some melancholy and unexpected catastrophe lays bare the feeble foundations on which they have been raised."

Long before the French disasters of 1870, the prohibitive battle losses of the American Civil War, the cult of the offensive that drenched French tactical and strategical thought in 1914— and that drenched French soil for the next four years with the blood of her soldiers—and the reliance upon the Maginot Line in

THE NATURE OF MILITARY HISTORY

1939–1940, Mitchell warned that military history could be exploited to document false doctrine. Later a German general felt compelled to observe: "It is well known that military history, when superficially studied, will furnish arguments in support of any theory or opinion."

I would agree. I suspect, too, that no other field of history is under such great pressure to serve some didactic function. Indeed, Thucydides intended his great history as an object lesson for the benefit of later generations. "I have written my work, not as an essay which is to win the applause of the moment, but as a possession for all time." Hoping that his work would be judged "useful by those inquirers who desire an exact knowledge of the past as an aid to the interpretation of the future," Thucydides maintained that in the course of human events the future must resemble the past, "if it does not reflect it."

We can learn something from Thucydides. His careful attention to detail illuminates the art of war of the Greeks, but more important, his profound understanding of human nature, and of the character of a democracy at war, is as instructive to the reader today as to the next generation of Athenians. Of the revolution in Corcyra in 427 B.C. Thucydides observes:

> thus every form of iniquity took root in the Hellenic countries by reason of the troubles. The ancient simplicity into which honor so largely entered was laughed down and disappeared; and society became divided into camps in which no man trusted his fellow. To put an end to this, there was neither promise to be depended upon, nor oath that could command respect; but all parties dwelling rather in their calculation upon the hopelessness of a permanent state of things, were more intent upon self defence than capable of confidence. In this contest the blunder wits were most successful. Apprehensive of their own deficiencies and of the cleverness of their antagonists, they feared to be worsted in debate and to be surprised by the combinations of their more versatile opponents, and so at once boldly had recourse to action; where their adversaries, arrogantly think-

ing that they should know in time, and that it was unnecessary to secure by action what policy afforded, often fell victims to their want of precaution. In the confusion into which life was now thrown in the cities, human nature, always rebelling governed in passion, above respect for justice, and the enemy of all superiority.

One can almost hear the apocalyptic voice of Eric Severeid as he described present conditions. Where? In South Vietnam? Perhaps even in the United States.

The generation of Englishmen who fought the First World War liked to compare their situation with that of the Athenians—two democracies, they were, and each a sea power, locked in mortal combat with the most militaristic and authoritarian land power in the world. But this comparison, while interesting (and also hopeful in its outcome) is less instructive than Thucydides' profound commentaries upon the attitudes and actions of his contemporaries.

"Capital," he reminds us, "it must be remembered, maintains a war more than forced contributions."

"Zeal is always at its height at the commencement of an undertaking."

"The sufferings which revolution entailed upon the cities were many and terrible, such as have occurred and always will occur, as long as the nature of mankind remains the same. . . . In peace and prosperity states and individuals have better sentiments."

"The strong did what they could, and the weak suffered what they must."

"War is not so much a matter of weapons as of money, for money furnishes the material for war. And this is especially true when a land power fights those whose strength is on the sea."

"The strength of an army lies in strict discipline and undeviating obedience to its officers."

And finally, ". . . it is the habit of mankind to entrust to careless hope what they long for, and to use sovereign reason to thrust aside what they do not fancy."

THE NATURE OF MILITARY HISTORY

Thucydides even reminds us that there is nothing particularly new—or valid—in the domino theory, as he quotes from Alcibiades' attempts to persuade the Spartans to reenter the war after the Truce of Nicias:

> That the states in Sicily must succumb if you do not help them, I will show. Although the Siciliots, with all their people and blockaded from the sea, will be unable to withstand the Athenian armament that is now there. But if Syracuse falls, danger . . . from that quarter will before long be upon you. None need therefore fancy that Sicily only is in question. Peloponnese will be so also, unless you speedily . . . send . . . troops.

He also suggests that negotiations for a peace, once hostilities have occurred, are always difficult and lacking in trust. Commenting upon the discussions between Corinth and Corcyra, he states: "The answer they got from Corinth was, that if they would withdraw their fleet and the barbarians from Epidamnus negotiation might be possible; but, while the town was still being besieged, going before arbitrators was out of the question. The Corcyraeans retorted that if Corinth would withdraw her troops . . . they would withdraw theirs. . . ."

Now this is the proper way to make use of military history. "One of the most important objects of education," the future Confederate General Joseph E. Johnston told a Congressional Committee on the eve of the Civil War, "is to give habits of judicious reading." He was not referring to the memorization of the minutiae of a campaign; he did not identify history only with facts, and military history exclusively with battles. For the value of history, and therefore of military history, is not in the facts it communicates or even in the principles it illustrates. The value of history is that it can provide fresh insight into the past and hence a better understanding of the present.

Yet students and soldiers alike frequently fall into the trap of looking for precise, hard lessons which, once mastered, will lead inevitably to success on the battlefield. Such individuals are prone to see some basic truth in every pattern, and something of significance in every similarity.

Several years ago an article appeared in a magazine of popular history under the captivating title "U.S. Grant Invades France." "It certainly is a fact," the author begins (and there is no evidence that he writes with tongue in cheek) "that among the 489 officers and 614 rated men comprising COSSAC (Chief of Staff, Supreme Allied Command) in 1944, a majority had studied the brilliant Vicksburg campaign. . . . And the final plan for the invasion . . . bears a remarkable resemblance, point by point, to the plan used by the Union army under Grant to cross the Mississippi River and capture the fortress city of Vicksburg."

Perhaps the unwary, like this author, would be overwhelmed by the superficial comparison between the two campaigns. Likening the English Channel to the Mississippi, this writer saw in the breakout from the Normandy beachhead and the subsequent race to Paris and on to the German frontier a movement strikingly similar in outline and theory to Grant's moves to Jackson and thence to Vicksburg, once he got his forces safely established on a "beachhead" south of the Confederate stronghold.

Well, such nonsense concerning the Normandy campaign belongs properly to leftover fare from some Civil War Round Table, but how many casual readers, lacking sophistication in military matters and unfamiliar with the workings of history, would accept such an outlandish theory simply because it would never occur to them to reject it? The only *fact* in the article is the statement that COSSAC comprised 489 officers and 614 men. Even if Dwight D. Eisenhower had been an avid reader of history, like George S. Patton, this is not the way military history is normally utilized. Although military history frequently does provide a mental matrix for the formation of judgment, rarely if ever can it offer such a tidy blueprint such as that described in Grant's alleged invasion of Normandy! Yet the legend persists that Erwin Rommel derived his inspiration for his African campaigns from a tour of the Shenandoah Valley and Gettysburg in 1939. From a superficial similarity in pattern, it is often reasoned, there must be a cause-and-effect relationship.

The Civil War has always had a special appeal for those who like to jump to adventurous conclusions. At the time it was fought, professional soldiers abroad were quick to note that the war differed in its character from anything they had experienced. There were no decisive battles that they could see; the cavalry seemed to have lost its punch on the battlefield; and artillery was relatively inefficient in most battles. To a soldier versed in the campaigns of Napoleon, these deviations could be explained away. Guns did not work very well in the thick wilderness; the soldiers lacked the training and discipline of regular troops; and so on. To the degree that European soldiers could see that the campaigns approximated accepted practice, they took the war seriously; but the few who did learn any meaningful lesson were a distinct minority. Aside from the technical interest, the Civil War therefore seemed to offer nothing to the French or German officer and little even to the British. For our part, we remain so intrigued by the war, and so generally ignorant of military trends in the half century between Leipzig and Gettysburg, that we see the war as the first of the modern conflicts, the forerunner of the trench stalemate in 1914–1918 and—if we are to believe some enthusiasts—the model for the German *blitzkrieg*. Anyone who will believe this is capable of believing anything.

The Civil War may serve as an illustration of the use—and abuse—of military history. In 1935 the future Field Marshal Earl Wavell encouraged a distinguished military historian and theorist, Captain B.H. Liddell Hart, to investigate what history might teach about the use of armored forces behind enemy lines. Liddell Hart, who several years previously had written a didactic biography of William Tecumseh Sherman, went directly to the Civil War and particularly to the Confederate cavalry operations against Union lines of supply and communications in the West. From his analysis he concluded:

AN ACADEMIC HISTORIAN'S POINT OF VIEW

1. "When acting in close co-operation with the army, the mobile arm proved ineffective in its offensive action." Anyone familiar with the British desert campaigns in 1941–1942 would testify to the general truth of this observation.

2. "When used independently, for strokes against the enemy's communications, the mobile arm was occasionally of great effect. . . . Long-range moves seem to have been more effective than close-range." After the fall of France in 1940 and the entrance of Rommel into North Africa, the truth of this observation also became apparent.

3. Finally, when discussing cutting enemy communications, Liddell Hart concluded: "In general, the nearer to the force that the cut is made, the more immediate the effect; the nearer to the base, the greater the effect. In either case, the effect is much greater and more quickly felt if made against a force that is in motion, and in course of carrying out an operation, than against a force that is stationary. . . . Thus, unless the natural obstacles are very severe, or the enemy has unusual independence of supplies from base, more success and effect is to be expected from cutting his communications as far back as possible. A further consideration is that while a stroke close in rear of the enemy force may have more effect on the minds of the enemy troops, a stroke far back tends to have more effect on the mind of the enemy commander."

So much for what at least one student learned from a detailed analysis of some campaign from the pages of history. The defect of this method is best illustrated by the German General Friedrich von Bernhardi, perhaps the foremost German military writer alive at the time of the First World War. Bernhardi was one of the conservatives who believed, as late as 1912, that it was still possible to attack an enemy successfully from the front despite the machine gun and other modern weapons. The official view was that envelopment in either tactics or strategy was far more likely to produce favorable results, and to convince his readers of a truth he already felt, Bernhardi turned not to the

THE NATURE OF MILITARY HISTORY

Vicksburg campaign, but to Grant's attack at Chattanooga. Here, he wrote, the Union general by a frontal attack successfully pierced the center of the hostile army. "A splendid victory was the result of this ingeniously planned battle."

The fact of the matter is that the battle of Chattanooga was never planned the way it was fought. Grant intended no frontal assault—which would be news to General Bernhardi, who possessed only a superficial knowledge of the operation—but planned instead to feint with Major General Joseph Hooker on one flank and to attack with Sherman's force along the other. The troops who ultimately stormed the center of the Confederate position atop Missionary Ridge were intended simply to pin down the Confederates. Bernhardi's factual outline of the battle is correct—or at least correct enough to catch the unwary reader. And Bernhardi himself was an accomplished historian, for as a younger officer he had labored for several years on the German General Staff history of the wars of Frederick the Great. An American student of the war would view this interpretation as a travesty; Grant's center attacked only when his flank attacks had stalled and the troops decided on their own to scramble up the hill. The attack succeeded, not because Bernhardi's theory was correct, but because the Confederates had dangerously weakened their forces there to meet the earlier threats to their flanks. Chattanooga was, in fact, the exception that proved the rule, and any theory that rests upon this particular case from history is sitting on a foundation of sand. Had Bernhardi read about Malvern Hill, Fredericksburg, Gettysburg, Cold Harbor, Franklin, and Nashville, he would have reached quite opposite conclusions. Again one is reminded of Clausewitz's biting observation, recorded some eighty years earlier: "From this sort of slovenly, frivolous treatment of history, a hundred false views and attempts at the construction of theories arise."

Obviously the difference between Liddell Hart and Bernhardi in this instance is that the one sought to learn by an objective analysis of a military operation from history; the other was anxious to illustrate the validity of some theory he already believed in.

Often a military historian attempts to do both. When Liddell Hart pondered the future use of the tank, back in the early 1920s, he probably learned more from a searching examination of the Mongol campaigns of the thirteenth century than from the most recent war. In addition to some general truths, he learned specifically from the Mongols the likely effects of certain tactics and organization for tanks. Perhaps this is best illustrated by the juxtaposition of successive sentences from two of his early writings, his description of Mongol tactics in 1925 and his suggestions for a pattern of mechanized warfare written two years later.

The battle formation of Genghis Khan, he wrote, was comprised of five ranks, the squadrons being separated by wide intervals. The troops in the two front ranks wore complete armor, with sword and lance, and their horses also were armored. In other words, they served as heavy tanks. "The three rear ranks wore no armour, and their weapons were the bow and the javelin." *The actual tank attack should be made by combined units of heavy and light tanks.*

From the rear ranks of the Mongols "were thrown out mounted skirmishers or light troops, who harassed the enemy as he advanced." *The light tanks would lead to pave the way by drawing the enemy's fire and testing his defence.* "Later, as the two forces drew near each other, the rear ranks advanced through the intervals in the front ranks, and poured a deadly hail of arrows and javelins on the enemy. Then when they had disorganized the enemy ranks, they retired into the intervals, and the front ranks charged to deliver the decisive blow. It was a perfect combination of fire and shock." Liddell Hart's conclusions in 1927 on mechanized columns held:

> If found to be weak, the light tanks would go through the enemy defences all out with the battle tanks on their heels. If strong, they would halt on any suitable closeup fire

position, thus turning themselves into a screen of minute pillboxes, stationary to ensure aimed fire, yet capable of instant change of position at need. Through this screen the heavier tanks would sweep, and the position of every antitank gun which opened against them would be smothered with a thick spray of aimed machine gunfire from the light tanks. . . . Once the battle tanks were through the first layer of antitank defence, the light tanks would race ahead, pass through them and repeat the process. Thus the tank attack would be an alternating process of movement and a compound process of fire. According to the hostile fire and the circumstances, the light tanks might either make direct for a chosen fire position, or, like the Mongol horse archers, race closer to the enemy before wheeling about and retiring a short distance to their covering fire-positions.

We see, then, history providing the theorist with his basic ideas. But after Liddell Hart's theories were mature, he turned to general history to confirm the lessons that he, for one, had learned from studying isolated campaigns. And naturally he found what he was looking for. The "strategy of indirect approach," on the basis of a broad and often sketchy survey of the decisive wars of history, seems to the casual reader of Liddell Hart to have operated successfully throughout history. In point of fact, "history" in this instance taught the author little that he did not already know. What it did do was to provide him with illustrations to demonstrate that he had found the military equivalent of the philosopher's stone.

Count von Schlieffen, the German Chief of Staff before the First World War, also illustrates this point. The problem he faced every night, while trying to sleep, was how to plan for a successful two-front war against France and Russia. Necessity dictated that he must overwhelm his enemies, one at a time; geography and the backward state of Russian society dictated that he should concentrate overwhelming strength against France. But how could he expect to win a decisive victory in the areas open to German invasion, where lay the Vosges Mountains and great fortresses at Belfort, Toul, and Verdun?

AN ACADEMIC HISTORIAN'S POINT OF VIEW

The answer was suggested by history—very ancient history. Reading the first volume of Hans Delbrück's *Geshichte der Kriegskunst,* Schlieffen was much struck by the battle of Cannae, where Hannibal's army, after giving way in the center, had won a convincing victory over the superior Roman forces by enveloping the enemy on both flanks. Envelopment was the key. Eventually the Schlieffen Plan evolved, which originally provided for a deliberate weakening of the German forces along the common frontier, to lure the French to attack in Lorraine, while the bulk of the German forces marched through Belgium to strike the French on their northern flank and pin the French forces into the pocket formed by the Swiss Alps and the Rhine. Having once discovered this key, Schlieffen used it to unlock the secrets of more recent great captains. Writing at length on the campaigns of Frederick the Great, Napoleon, and Moltke, he naturally saw every victory as the result of an attempt to envelop the enemy. Against this massive array of evidence, no wonder Bernhardi was scrambling for historical support for his own pet theory.

There are other uses—and abuses—of military history. One should realize that while some books represent research undertaken to learn, others are written as devices to teach. This was particularly true of the nineteenth century, when texts like Sir Patrick MacDougall's *The Theory of War,* Sir Edward Bruce Hamley's *The Principles of War Explained and Illustrated,* and Matthew Forney Steele's *American Campaigns* made use of history to illustrate the theories of others. Steele's volumes are a good example of this kind of literature. Steele was a major when he wrote the book at Fort Leavenworth, and it endured as a military text at West Point for two generations. His comments on strategy, however, resemble a tossed salad, for he sprinkled his pages with appropriate quotations from most of the leading military writers of the nineteenth century. This explains why he could not make up his mind whether a legitimate objective of strategy should be a city, the destruction of the enemy army, or a political objective; why Richmond, *as well as Lee's army,* were valid Union objectives in 1862, but *Lee's army alone* was the

31

THE NATURE OF MILITARY HISTORY

true objective two years later; why on one page he accepts Colmar von der Goltz's contention that "the immediate objective against which all our efforts must be directed is the hostile main army," whilst on another he states blandly—and blindly!—"It is as much the province of strategy to dishearten the hostile people . . . as it is to defeat and destroy their armies."

In books such as these, the standard fare for military students before World War II, it is not history that teaches, but a teacher using history to illustrate accepted principles. It is theory that provides the basis for historical judgment upon the strategy and tactics of American campaigns. And if no one set of principles emerges—since each chapter illustrates something different—it is only because the American army itself had no distinct doctrine at the time.

If you were to look at the German histories of the Boer War or the Russo-Japanese War, you would find the same lessons consistently driven home by every campaign. German histories never failed to teach the soundness of the German conception of war, and the triumph of the Japanese in Manchuria was regarded by German critics with much the same satisfaction as a teacher feels contemplating the achievement of a prize pupil. For the Japanese Army had been German trained, and what historian—particularly if he was a serving officer at that time—is apt to render historical judgments that fly in the face of accepted and official doctrine? Today, in our Army and particulary in the wake of an unpopular and unsuccessful war, I would expect officers to feel fairly free to criticize policies as well as doctrine. It is expecting too much of human nature, however, to look for much deviation in the days when Europe was divided into two armed camps, containing soldiers who themselves had only limited experience with war, and when intense nationalism was widely recognized as a force for good nearly as worthy as Christianity and the White Man's Burden.

When called upon to write official history, particulary the official account of some campaign in which their own army was involved, authors have found different reasons for abusing history. The British official history of the military operations of the First World War often represents a distortion, not necessarily of fact, organization, or the description of battles, but rather in interpretation. "The facts are all correct," an eminent theorist wrote of the early volumes, "but the atmosphere of reality is completely wanting." Brigadier James E. Edmonds, the author of most of the volumes and editor of the rest, set out originally to "discover what actually happened, in order that there may be material for study, and that lessons for future guidance may be deduced." Privately, he admitted: "I want the young of the Army who are to occupy the high places later on to see the mistakes of their predecessors, yet without telling the public too much." Later he confessed: "One can't tell the truth"—old army loyalties too often got in the way, and as General Edmonds approached his final volumes, he allowed others to write the body of the texts, while in his editing chores he "twisted the narrative to the outlook of British Headquarters at the time." One of his colleagues did produce an outspoken volume on the Gallipoli campaign, but apparently there were things that he too did not dare to tell. When asked by a publisher in 1935 to write a short personal account of the campaign, Brigadier Aspinall Oglander replied candidly: "I could not write an unofficial account without expressing personal convictions which would be in contradiction of many of my statements in the official volume, and I feel that this, in addition to being rather undignified, and a lapse of taste, would be quite unfair to the Government, which paid me for the official history."

Small wonder that General Sir Ian Hamilton, the British commander at Gallipoli, would write of an earlier war: "On the actual day of battle naked truths may be picked up for the asking; but the following morning they have already begun to get into their uniforms."

This is not to say that this problem existed in any of the official histories produced by our own country after World War II. According to Martin Blumenson, author of a recent biography of Patton: "The response of the historical profession to the volumes in the official series seems to prove that the historians in the employ of the Army are writing honest history." Pointing to the three factors that he considered responsible for this judgment—the use of qualified, professional historians; the free access granted to all the records; and the lack of any censorship, Blumenson concluded: "Our findings . . . are anything but official. Because of the conditions of our work, if we do not write honest history, we have no one to blame but ourselves." And a British historian, not known for his friendly views of his American allies, testified: "The attitude of the American authorities to historical inquiry is not only beyond all praise, but also not the least of the hopes of mankind in a wider sphere. They are not afraid of the facts of history."

There is a focus to these rambling remarks, and also a point. Military history is prey to problems and pressures that are involved far less often in the writing of other kinds of history. The theorist, anxious to find some idea that he can apply to a problem that concerns him, can search the pages of wars centuries old and continue to find gold nuggets. Often the theorist who finds a convincing answer is then anxious to use history to help convince others; this too can lead to distortion. Those who seek to find patterns in war often approach history from the wrong direction. Knowing already what they want to find, they tailor the facts of history to fit their preconceived notions. This is not necessarily wrong, nor does this method lead necessarily to false views. But it is not history.

Soldiers who take up the pen find it difficult, for a variety of understandable reasons, to forget that their hand is also expected to wield the sword. There are reputations to protect, old loyalties that cannot be forgotten, governments that must be satisfied. Too, there is the civilian who is against war and seeks to discredit those who must fight it. Mr. Fair's recent book on the stupidities of the not-so-great captains would fall in this category.[1] Many civilian historians—myself included—lack the

[1]Charles Fair, *From the Jaws of Victory: A History of the Character, Causes and Consequences of Military Stupidity, From Crassus to Johnson and Westmoreland* (New York: Simon & Schuster, 1971).

personal experience or technical knowledge to write certain kinds of military history. I can follow the movements of brigades and divisions through a battle, but never having fired a gun in anger myself, I can only depend upon others for the psychological dimensions of battle. Nor am I sufficiently at home with modern weaponry to write anything worth reading on that subject. This does not stop many military historians, however, from taking on such subjects, because the material is popular, and full of implications for us all. We are frequently asked questions about the current situation that would stump an Assistant Secretary of Defense.

One could mention other problems connected with the writing of military history. It is still a young subject, and one that is often in the hands of immature men. It is eaily distorted by national pride, excess enthusiasm, and strong prejudice. It can claim more than any form of history deserves; and it seems to be inevitably connected with "lessons," as if the lessons of history are expected to stand apart from the historian and independent from a period. One can write of social movement, of intellectual currents, of political events, or of diplomatic affairs, without being expected to deduce lessons that will be applied by society or by the government.

In our world, and particularly in our society, much—indeed everything—depends upon an electorate capable of understanding the issues and rendering sound judgment. History can help us all perceive things with greater insight and understanding. It follows, then, that because so many of our problems today involve the use of the military instrument in some way, military history is relevant—fully as much to the civilian as to the soldier.

But it requires a fairly sophisticated person to separate the wheat from the chaff. At least he must understand that not all that he reads from books—or newspapers—can be taken at face value. He must search for his own answers, and he must learn something about evaluating what he is reading. The basic problem in communications today is probably not what the government may or may not be concealing from us, it is how to understand the varied, detailed, and often contradictory information that is made available.

Military history can be of increased use. But it can also be abused, and one would do well to recall that old aphorism of a German writer 200 years ago: "A book [even a good book] is a mirror: when a monkey looks in, no apostle can look out."

II

Armed Forces and Society

Armed Forces and Society:
Some Hypotheses
THEODORE ROPP
page 38

Cultural Change,
Technological Development
and the Conduct of War
in the Seventeenth Century
DANIEL R. BEAVER
page 73

===== COMMENTS =====

The "new military history" developed by civilian historians both in America and abroad since World War II has dwelt especially upon the interactions between armed forces and the societies they serve. If any one theme has been central to the "new military history," this is it—so much so that it has become a commonplace to observe that the first key to understanding any armed force is that it is an extension of the society that created it. A corollary to this view is that an armed force is likely to be no stronger than the society it serves; a sick society can hardly produce healthy armed forces.

Thus the British military historian Corelli Barnett—an exceptionally gifted member of the outstanding post-World War II generation of military historians in a nation that has consistently produced superior practitioners in this field— made his book **The Swordbearers: Supreme Command in the First World War** *(paperback edition, Bloomington: Indiana University Press, 1975) a study of how four notable World War I military commanders embodied the strengths and weaknesses of their nation-states. Barnett's "Sailor with a Flawed Cutlass," Admiral Sir John R. Jellicoe, had to fight with a weapon that was flawed—the Royal Navy of 1914–1918 with its weaknesses in ship design, armament, and armor—because the British Navy was an extension of British society, and the failings inherent in the socially stratified society of Victorian Britain were by the time of the Great War inevitably showing up in the Navy. "There was thus nothing accidental, nothing of bad luck, nothing of blame on individual officers," writes Barnett, in what went wrong for Jellicoe and the British Navy in the battle of Jutland. Rather, Jutland "was part of a vast process of dissolution that began about 1870, when the British forgot that life is a continued response, with daily new beginnings; that nothing is permanent but what is dead." (p. 188). "In distant retrospect," Barnett* concludes, *"Jutland was one of the critical battles of history; it marked the opening of that final phase of British world power and maritime supremacy that was to end in 1945, with the British battle fleet no more than 'Task Force 77' in the United States Pacific Fleet, and Britain herself reduced to financial dependency." (p. 189).*

This is dramatic prose and perhaps brilliant analysis; but it is also a bit too neat. It is reminiscent of Arnold Toynbee's writing off the Byzantine Empire as dead at about the year 1000, though in fact the empire had four and a half more centuries of existence ahead of it, which Toynbee blandly dismisses as fossildom—all the more reminiscent because while the Byzantine Empire undoubtedly suffered numerous illnesses during the period when Toynbee pretends it was dead, that sick society managed until almost the end to project generally impressive military power. If armed forces are merely extensions of the societies they serve, if sick societies cannot project healthy armed forces, then military historians are in the predicament of Arnold Toynbee, with too many centuries and too many experiences of history to explain away.

We are back to the messiness of the meanings of history. To recognize armed forces as being extensions of their societies is surely a useful perception; to have set military history into the broad context of social history, rescuing it from an almost exclusive focus on the battlefield, is one of the most considerable achievements of the new social history. But in history, every generalization is dangerous; the dictum that armed forces are extensions of societies is another historical generalization that requires the most diligent caution in applying it lest it merely confound and confuse.

In 1971 Theodore Ropp essayed for the New Dimensions course a more subtle examination of the interaction between

═══════════════════════ COMMENTS ═══════════════════════

armed forces and societies, using as a point of departure Stanislav Andreski's also subtle Military Organization and Society *(revised ed., Berkeley: University of California Press, 1968). If the commonplace dictum about armed forces reflecting societies is too simple, perhaps here we meet the beginning of the opposite problem, of the subtleties getting out of hand; but the reader should know that to perplex as a challenge to thought is part of Professor Ropp's style. Here we traverse a substantial part of the whole terrain of world military history with an author whose* War in the Modern World *(revised ed., New York: Collier Books, 1962) has long since been accepted as a classic synthesis of the subject; as a contributor to Edward Mead Earle's* Makers of Modern Strategy: Military Thought from Machiavelli to Hitler *(Princeton: Princeton University Press, 1941), Professor Ropp has also been part of the "new military history" since the volume that foreshadowed it. At Duke University he has probably trained more graduate students in military history than any other mentor in the United States. In 1972–1973 he was the first holder of the visiting professorship of military history at Carlisle Barracks.*

Armed Forces And Society: Some Hypotheses

THEODORE ROPP

THIS TOPIC would be impossibly broad except for adding "dimensions of depth and historical experience" to one's existing "store of knowledge and concepts." One uses history for mind expanding, not puzzle solving; whether it becomes a mind-expanding drug depends on the user. What Sir Keith Hancock calls "funded experience" is essential in social puzzle solving; but tunnel vision, historically based strategies and tactics have been disastrous. On wider topics, unfortunately, leaders have become even drunker on history. For while "it is possible," as Benedetto Croce noted, "to reduce to general concepts the particular factors of reality . . . in history, . . . it is not possible to work up into general concepts the single concrete whole."[1] George S. Patton's choosing to paraphrase Napoleon: "To be a successful soldier you must know history," instead of using Polybius' "History is the truest education . . . for political action,"[2] indicated this problem. For Patton was one kind of leader who is essential in any army, one who, like Charles XII of Sweden or Horatio Nelson, was drunk on fighting; while Patton's ally, Viscount Montgomery, was another essential kind of man, a cautious charismatic. Neither would have been a good supreme commander in an Allied peoples' war in which Winston Churchill and Franklin D. Roosevelt were great popular leaders. But Churchill, like Charles de Gaulle, often drank too much history. Roosevelt, like Dwight D. Eisenhower, knew little more about it than they all knew about economics.

A common sense feeling for men, tactics, and technology is our requirement for leading "armed forces . . . [which reflect] the social and cultural entities they have been created to serve, and . . . strengths and weaknesses . . . of the nations they represent." But since state, society, culture, and nation did not—after the fall of Rome—again become even approximately synonymous in the West—or in westernizing Russia—until the peoples' wars that began with the American and French Revolutions, let me apply my own Military participation, military Subordination, and social—instead of military—Cohesion variant of Stanislav Andreski's MSC taxonomy for the relations of *Military Organization and Society*,[3] chiefly to our era, after brief comments on the Romans and on Western feudal warfare.

The dangers of such models of historical reality have already been noted. Their value, like that of a formal staff estimate of a situation, is that they force us to look at every aspect of what Karl von Clausewitz called war's "chameleon, . . . [which] in each concrete case changes somewhat its character . . . of the original violence of its essence, . . . the play of probabilities and chance, . . . [and] of the subordinate character of a political tool, through which it belongs to the province of pure intelligence."[4]

In one of his few formulas Clausewitz noted that, "If we want to overthrow our opponent, we must proportion our effort to his powers of resistance, . . . expressed as a product of two inseparable factors: *the extent of the means at his disposal and the strength of his will."* The first could be estimated, "as it rests (though not entirely) on figures"; the second was "only approximately to be measured by the strength of the motive behind it," since "Bonaparte, war, through being first on one side, then on the other, again an affair of the whole nation, . . . had no visible limit . . . . The cause was the participation of the people in this great affair of state, . . . [arising partly from] the French Revolution, . . . partly from the threatening attitude of the French toward all nations."[5] Clausewitz and other Prussian conservative reformers, feared that peoples' wars might lead to the people's participation in other "great affairs."

SOME HYPOTHESES

The major nineteenth-century political achievement of these conservative reformers was to be their co-optation of German nationalism, liberalism, and the industrial revolution. But the last—which was to revolutionize the relations of "technology, culture, and warfare," and to create new internal, international, and technical military problems—was still a distant British phenomenon in Clausewitz's day.

So "Armed Forces and Society" touches on so many subjects that I will begin with the hardly original ideas that (1) in dealing with large areas and/or long periods of time, all historical classifications and general theories have major weaknesses; that (2) as the MPR—Military Population Ratio—increases, viable armed forces more clearly mirror states, cultures, societies, and nations; and that (3) technology was a less easily recognizable and deliberately manageable variant in military systems before the twentieth century. I will conclude with (4) a set of propositions or hypotheses on the historically schizophrenic relations of American armed forces and society, to illustrate some of the strengths and weaknesses of the most familiar study of "armed forces . . . as an important aspect of American politics," Samuel P. Huntington's *The Soldier and the State*.[6]

The Romans

Rome has been an almost irresistible source of moralization about armed forces and society whenever moral, political, or military decay seem to explain defeat. For the Romans reversed modern Western experience by going from Andreski's MSC (capitals show high MPR, military subordination, and military cohesion) "widely conscriptive" republican system—with its mythic Horatius and Cincinnatus—to an mSC "restricted professional" one during the social and civil wars and the empire that allegedly ended in the West in 476 A.D.

The Emperor Justinian (527–565) and his successors regained Carthaginian Africa, Spain, and Italy and destroyed Sassanid Persia, at social costs which help to explain the rise of Islam in the next century. The Byzantine system became MSC until the Arabs were finally repulsed from Constantinople (717–718), before going into the West Roman pattern of "decline" to mSC and Andreski's lower case msc "feudal" system.[7] While Andreski's model is most illuminating, changing his C from military to social cohesion makes it possible to use Sc combinations that are militarily "impossible"—high military subordination cannot go with low military cohesion—while high military subordination and low social cohesion may explain the successes of some feudal armies. And the same revision may help to explain Rome's "progress" from MSC to the end of the Second Punic War (218–201 B.C.), to mSC professional for overseas service, MSc professional during the civil wars, and an Augustan mSc military and political reorganization (31 B.C.–14 A.D.), which brought peace and mSC social consensus.

Later praetorian coups and civil wars, expansion, and barbarian infiltrations and invasions were less important than the growing social cohesion which ended with the triumph of Christianity. In "government and law," Chester G. Starr, Jr., remarks, "the purely Roman contribution looms the largest; in its sympathetic attitude toward men—humanitarianism—the Empire broke its freshest ground, . . . above all in the body of Christian teaching and practice." While MSc and mSc systems are inherently unstable and this revision of Andreski's categories moves his "Trekboers" and "North American frontiersmen" from Msc or "inarticulate sub-tribal" to MsC or Cossack "tribal,"[9] these new classifications may also help to explain what happened when the Prussian and Russian msc feudal forces went to mSc and MSc in which—as Clausewitz had feared and as Sir John Monash noted in 1920—Prussia's "proverbial 'iron discipline . . . ultimately broke down completely under the test of a great war.' "[10] The National Socialist German Workers Party and army were MSC; its pressures and Communist industrialization made a shaky MSc Red army into an MSC opponent.

Clausewitz, who was no classicist, made only one reference in *On War* to the Rome which so fascinated eighteenth-century republicans. "She became great more through the alliances which amalgamated with her, . . . than through actual conquests. . . . Only after she had spread . . . over Southern Italy, . . . she began to advance as a really conquering power . . . . Her forces were immense . . . . She no longer resembles the ancient republics . . . . She stands alone."[11] With her allies and colonies—most commonly of Romans who got conquered lands but lost their Roman citizenship while retaining the private rights of allied and Roman citizens—Rome could call on 750,000 men during the Second Punic War. Her twenty to twenty-three legions and attached allied forces numbered some 250,000 men from a population of about 3,750,000, not too far from modern standards of 10 percent of the total for a people's war. Whether the parallelism is conscious or not, Clausewitz's explanation for Rome's success is Polybian. "The Romans had quite adequate grounds for forming their plan of world rule, . . . as well as ample resources for attaining this end."[12]

With Herodotus, Polybius was the ancient historian who paid the most attention to technology. He noted the careful workmanship of Roman weapons, how the point of the no-return javelin was made to bend when it hit, and the solidity of the iron-edged, iron-bossed Roman shield against heavy Celtic swords or axes. And the speed with which the Romans adopted the Greek heavy cavalry spear when they had to fight such cavalry was further evidence to him that "the Romans . . . above all others are good at changing their ways and imitating better practice."[13] We know that they borrowed superior barbarian ironworking techniques and that their farmers were working newer and less eroded and malarial lands than those of their Carthaginian and Greek rivals. Nightly fortification, careful scouting, and good march discipline enabled the Romans to march straight through enemy country to its strongpoints. The overconfidence which led them into occasional ambushes caused some naval disasters after their first victories during the First Punic War.

The Romans use force to accomplish everything, and consider that they must necessarily achieve, at any cost, whatever they plan . . . . [They often] succeed because of this determination, . . . [but when] they clash with the sea and atmosphere, they meet with great disasters. This . . . will happen to them, . . . until they cure themselves of that reckless belief in force that makes them think that, for them, every season must be a fit one for sailing and traveling.[14]

The older Carthaginian constitution was past "its climax, while the Roman constitution attained its climax, at that precise moment. Consequently the common people of Carthage had already taken over the greatest share of power, . . . while at Rome the Senate held the supremacy." But Polybius did not stress this Greek cyclical model, or the idea that "Italians are naturally superior to [possibly underfed] Phoenicians and North Africans, both in bodily strength and personal bravery." He stressed the "customs" which developed "these qualities among their young men" and "a common sense felt by the whole people . . . . Superstition . . . is dramatized . . . to the highest possible degree . . . . [In] a state consisting entirely of learned men, . . . [this might] not be necessary, but since every populace is unstable, . . . the only recourse is to restrain the masses by means of dark fears and this kind of ceremonial."[15] As a Greek outsider, Polybius (c. 201–121 B.C.) did not have to write omen-mongering history, but he would have seen its importance for Livy (59 B.C.–17 A.D.), who was preaching the ancient civic virtues after the long civil wars that had ended in the Augustan Principate.

Though this cannot be divorced from the fact that Polybius' welcome by the Scipionic circle made him our primary source for the events he described, his analysis of Rome's rise is quite close to our own. Starr sees Roman mythic history reflecting the realities;

> of a simple, conservative agricultural society resting on a long tradition of family life. Beset by enemies, . . . this society had to fight hard and continuously: . . . *virtus*

SOME HYPOTHESES

... meant pre-eminently military prowess .... Expansion abroad required unity ... [and] softened internal problems by giving foreign outlets; ... the military needs of the state forced the aristocrats to give way to the commoners at crucial points .... The Romans won by ... gaining alliances with some possible foes ... and carving up actual opponents by driving wedges through their territories, ... [building] roads, ... [and] establishing colonies ... at key points; ... conquered states on the average lost one-third of their land ... [to] Roman settlers. Otherwise the defeated states were not unduly penalized .... They agreed to furnish a set number of men to the Roman army, but they paid no taxes and retained [local] autonomy .... Particularly impressive was Rome's willingness to grant its own citizenship to subject states .... [Those] of Italy were much less fully developed than the cities Athens had tried to dragoon into its Aegean empire, ... [but] Roman liberality must also be given credit .... The sensible, practical, tenacious character of the Romans goes far toward explaining not only the initial conquest but also the relatively easy acceptance of their rule.[16]

More distant wars, slave revolts, urban and allied unrest, and defeats in Gaul led to the formal professionalization of the Roman army (105 B.C.) by Gaius Marius. Men were conscripted or enlisted for up to sixteen years. Generals who could get their veterans land also let them keep their citizenship and gave it to allied veterans at the end of their service. Their victories brought in tribute, land, and slaves. Military professionalization, civil wars among the oligarchs, economic ruin for the peasants, and prosperity for the plantation owners were contemporaneous. The primary factor, Andreski notes;

> is the strength of the [soldier's] links ... to the civilian society. ... [Their] alienation may be the result of long-term service, or of foreign origins, or of ... recruitment ... [of] a special psychological type, or of stationment in distant lands or ... [in an alien] cultural environment. ...

In Ancient Rome military revolts began when the army acquired these characteristics.

The Roman peasant recruit may have been "ignorant or politically apathetic";[17] but he hated barbarians, foreigners, and the barbarian and foreign slaves whose exploitation was one cause of his poverty. To outsiders the Romans must have looked liked irresistible army ants, swarming from centrally located Italian hills. They took time off to fight each other or their slaves, or to build new paths and fortified hills, but they always came back for more land, slaves, and plunder. So both barbarian and civilized still had "an enemy that could be feared and envied while still despised, conducive to comforting feelings of national or tribal solidarity, and . . . cathartic outbursts of warlike zeal from either side when the pressures of passion or policy demanded it."[18]

Augustus reduced the Roman army to twenty-five legions, some 300,000 men with auxiliaries, not much larger than that of the Second Punic War which had been drawn from less than a tenth as much population. This force, now permanently settled near the frontiers, had grown to 350,000–400,000 by the time of Marcus Aurelius (161–180). By that of Constantine (324–337, the first Christian Emperor), it consisted of field forces of 200,000 men and local forces of around 350,000. The empire's population has been put at 50–70 million under Augustus, 50–100 million in 250, and 70 million under Constantine. There were some recruiting shortages, but evidence of a general decline in standards is unconvincing, except to indicate a greater number of Easterners as the Western lands were exhausted by a Mediterranean farming system that had reached and passed its soil and climatic limits.[19] Tribesmen pushed in or were allowed to settle lands depopulated by plague or soil exhaustion. After the army was settled on the frontiers, its barbarization and the barbarians' Romanization were pretty inevitable. The frontier was eventually stabilized along the line of the Rhine and Danube to Romania. After Augustus had lost three legions in the Teutoburger Forest in 9 A.D. and his successors had begun to penetrate the steppes north of the Black Sea, Rome ran out of easily conquered lands and slaves. The *limes* were fronted by generally friendly and dependent Romanized tribesmen and backed by roads, Black Sea and English Channel naval squadrons, river patrols, and more cavalry and mounted infantry.

While outside pressures increased against the still viable East, the resistance of the economically declining West decreased. But Rome's professional forces retained their flexibility and fighting ability, refortified many cities, and built the fortress complex of Constantinople, nearer the sources of heavily armored steppe cavalrymen with saddles, stirrups, heavy Persian horses, bows, swords, and lances. An East Roman defeat near Adrianople in 378 was followed by the almost immediate development of similar East Roman cataphract cavalry. Invading "hordes"—only 80,000 for the whole Vandal tribe—were shouldered westward into Gaul, Spain, and Carthage, and were invited to Rome, which they sacked in 455, by the Western Emperor Valentinian III's widow. In the meantime Attila's Huns had been defeated or bought off from Italy by Pope Leo I. Valentinian had then murdered Aetius with his own hand, and had been murdered by two of Aetius' bodyguards. But these Gibbonian events and the deposition of the last Western emperor by another barbarian chief in 476 had only set the stage for Justinian's recovery of the parts of the earlier Roman republic which could be most easily reached by his sea and river-borne heavy cavalry. Their overall numbers are hard to determine. Archibald R. Lewis' *Naval Power and Trade in the Mediterranean, A.D. 500–1000,* suggests that 15,000 men recovered Vandal North Africa, that only 25,000 to 30,000 overthrew the stronger Ostrogothic state in Italy, and that forces which "cannot have been large" recovered still more distant Andalusia.[20]

Localized Medieval Warfare

Walter Goffart's brilliant "Zosimus, The First Historian of Rome's Fall" notes that "once the history of the Roman Empire is written as a story of decline, the range of variation becomes quite narrow."[21] Edward Gibbon's twelve centuries from Marcus Aurelius to the fall of Constantinople in 1453 gave him many variations on a few themes. He is weakest on the age of Justinian, on the Roman armed forces' recuperative powers, and on the ways that Polybius' "superstition" became Christian "fanaticism" to aid social cohesion. Since the comparative decline of the United States may bring out new Gibbonian· Early Christian attacks on the whole of American society, or draw equally Gibbonian defenses of our mid-century popular mass forces from those who fear a return to our early twentieth-century professional forces, we may note that our comparative decline may be somewhat analogous to one factor in that of Rome. The adoption of "American" technology by new and old Avis "we try harder" states is narrowing the technological gap between them and us as it was once narrowed between the Romanizing outsiders and a technologically mature Mediterranean civilization.

The economic decline of the West was not mythical, not wholly due to Roman unwillingness to invest in laborsaving devices[22] rather than in land, public works, and fortifications. The Mediterranean world lacked water power and animal fodder. Both became exploitable, as Lynn White, Jr., sees it, only after a new three-field farming system and the concept of a power technology had developed in medieval northwestern Europe.[23] Alfred Thayer Mahan's new-mercantilistic theory of sea power was to stress the connections among bulk maritime trade, labor, overseas colonies, and navies.[24] The peace which had brought prosperity to the then new Western lands began to cut into Roman bulk trade as soon as Rome's colonists could supply themselves with local products. The metal trades were not in bulky ores and fuel, and no new crops built bulk trade as Atlantic fish, slaves, tobacco, sugar, and timber did in the early modern era. Pliny the Elder (*c.* 23–79) noted that seamen were already underemployed and that the Mediterranean was being fished out, but the main reason for the decline of Roman trade in the West was the localization and subsequent decline of agri-

SOME HYPOTHESES

culture and manufacturing. And the Egyptian grain ships which fed the Roman proletariat from the Emperor's personal farm were redirected to Constantinople as Rome lost population.

The warm wet climate which had pushed the Mediterranean system of farming beyond its normal geographical limits aided erosion, silting, and malaria in the West, and increased barbarian population pressures. The Arabs and Vikings took over what was left of Western maritime trade in the ninth century. The Westerners used the Po, Rhone, and Rhine rivers, with pack horses, men, and wheelbarrows crossing the Alps, during this mild climate cycle. In the tenth and eleventh centuries Christianized ex-Vikings or Normans then helped Italian traders to win command of the Western Mediterranean, Sicily, and the routes that carried Crusaders to the Holy Land, after the Byzantines had lost Anatolia, their last major land recruiting base, to the Seljuk Turks after the battle of Manzikert in 1071.

There is no doubt that the military systems of the Western Early and High Middle Ages—500–1300, before the crossbow, longbow, trebuchet, and other high-angle stone-throwing machines, and gunpowder began to end the dominance of the armored horseman and his castle—rested on localized agrarian economies. Fortification is always a way of investing seasonally underemployed agrarian labor in armaments. And if there is some doubt about when White's lancers took to lancing, there are equal doubts about some feudal systems as msc, or mSc, or even mSC during external and internal Crusades. The two latter formulas may help to explain the successes of the Carolingian, Norman, Byzantine, and Crusader feudal armies, all of which had a considerable degree of military subordination, as well as effective logistical and engineering support systems.

Here, to take only one example, John Beeler's account of the Norman conquest of England in 1066 is most convincing. William of Normandy had gathered adventurous lords and mercenaries from as far away as southern Italy in what might be called an mSc anticipatory professional army. He landed them from 450–700 ships the same September that Harald Haardrada,

King of Norway, who had once fought the Normans in Sicily as a Byzantine mercenary, landed in northern England. The English King Harold defeated and killed Harald at Stamford Bridge on September 25, after the latter had defeated a local force on the 20th, and marched 250 miles in two weeks with his housecarl infantry after he learned that William had landed on the 28th. The latter had thrown up the wooden castle that can be seen in the Bayeux tapestry, and had plundered the country around Hastings. Beeler, the recent editor of Charles W.C. Oman's classic *Art of War in the Middle Ages,* then shows that the resulting battle was not one of Oman's examples of " 'a schuffle and scramble over a convenient heath or hillside.' "[25] William had about 3,000 cavalry who could also fight dismounted, 4,000 infantry, and 1,000 archers; Harold some 2,000 housecarls and 6,000 militia, a few archers, and no cavalry.

Both forces showed considerable military subordination and cohesion. William won the battle by a frontal assault after an allegedly "feigned flight" to draw the English off their hill. He took Dover without a siege, marched around London to cut off its supplies, and convinced "the English that resistance was futile." Anglo-Saxon England had few fortifications. The Normans covered it with quickly built motte-and-bailey castles which were the successors of Roman cantonments in occupied areas. The conclusions of Beeler's survey of Anglo-Norman England until 1189, when Richard I set out on the Third Crusade, are that there was then no "knightly monopoly of military service" and that "the combination of horse and foot [the chain mail tunic, kite-shaped shield, and conical cap were light enough for either kind of fighting], of archers, spearmen, and lancers, was far more flexible . . . [and adaptable] to terrain and enemy dispositions, than . . . formations composed entirely . . . of heavy cavalry."[26] Either type of force could be feudal, composed of professionals who owed military service for land tenure.

The reasons for the decline of the medieval knight are more familiar. More powerful missiles led to the development of plate

SOME HYPOTHESES

armor, which so blinded horses and men that they might pile up in the simplest ditch. Full armor was too heavy too allow dismounted men to get back on their horses. While kings crusaded, fortified towns, castles, and notions of feudal honor sprouted. The Teutonic knights' assault on the tired Poles at Tannenberg in 1410 paused for a one-man charge at the Polish king.[27] Far more important, the fourteenth-century storms, drought, floods, and plagues which opened a Little Ice Age that lasted until the mid-nineteenth century, depopulated whole villages from China to Middle America, also raised the Western European demand for labor and labor-saving devices, and aided the spread of many inventions and borrowings from Islam and, after the Mongols had reopened direct European contacts with the Far East, from China. Many of these devices, adaptations, and inventions are familiar. The lateen sail, the windmill, the compass, the pound lock gate which helped to create the modern Netherlands and its maze of water defenses, gunpowder, and the printing press helped to increase the wealth of fortified trading and manufacturing towns and enabled some new monarchs to recruit new mercenary standing armies to drive out foreign invaders and to knock down offending castles. Some of these new monarchs were supported by a growing social and national consensus of churchmen, traders, and artisans against the feuding nobles' exactions, although these were not profitable enough to keep poverty from forcing many of the nobles to take service as royal or city mercenaries. But the historical moral of this tale is that the Middle Ages cover so long a period of time and such diverse areas that almost any generalization can be supported from some of our mind-expanding models.

Three Great War Commanders

Partly because Continental Europe's nations-in-arms of 1914 had been so deliberately organized for a Great Peoples' War, their military and political leaders were to be bitterly criticized for not delivering the promised Napoleonic and Moltkean quick victories. Most historians are still under the influence of those critics. The most sensible view is still Georges Clemenceau's maxim that "War is too important to be left to the generals." Giulio Douhet charged that the generals were not only wrong on *"the formidable efficacy of firearms,"* but also that they had become a self-centered clique of professionals "competent only by definition" and "out of touch with the living, acting operating nation."[28] My example of the World War I professional general, Douglas Haig, was charged by David Lloyd George, the first man of the people to become British Prime Minister, with ignoring the talents of such militiamen as my two examples of the nonprofessional general, Arthur Currie and John Monash, and with being one of "two or three individuals who would rather the million perish than that they as leaders should own—even to themselves—that they were blunderers,"[29] and Currie was to be charged with deliberately slaughtering Canadians to satisfy the vanity of British higher commanders.[30]

A rather similar argument had erupted in the 1890s. Henri Philippe Pétain had not been promoted before 1914 because he was dubious about the French General Staff's doctrine of the *offensive à outrance* against the new smokeless powder fire weapons. Colmar von der Goltz, the author of *The Nation in Arms*, later claimed that he had not been made Chief of the German Army's General Staff because his defensive ideas might weaken its offensive spirit.[31] The whole discussion, which had spread to the Anglo-Saxons, had led to new economic arguments against war and to some new hypotheses—to be noted in connection with American views of the relations of armed forces and society—about how industrialization might be affecting people's views of war.

John Terraine's *Educated Soldier*,[32] Douglas Haig, fifty-three years old in 1914, had spent three years at Oxford before graduating from Sandhurst. He became commander of the British Expeditionary Force in December, 1915. Currie, thirty-

SOME HYPOTHESES

nine, Starthcona Collegiate Institute, was a British Columbia realtor and militia artilleryman. He had just completed a staff course in 1914, when he got an infantry battalion. He commanded a division during the 1916 Somme battles, and got the Canadian Corps before its participation in those summer and fall offensives of 1917 to which failure has given the generic name of Passchendaele. Monash, forty-nine, had civil engineering, arts, and law degrees from Melbourne. He had been a militiaman for nearly thirty years before getting an Australian infantry brigade, a division in July, 1916, and the Australian Corps in May, 1918. None of these men had much charisma. Haig was shy and inarticulate, Currie large and worried by his prewar debts, and Monash too businesslike in applying his knowledge to what he saw as the business of war.

Haig, a cavalryman, was the slowest to grasp the new military technology. Intemperate as some of Lloyd George's charges against him were, Haig was the prisoner of a system in which high commanders did not have "to view for themselves something . . . of the terrain of attack and the nature of the operations they were ordering their officers and men to undertake."[33] Terraine's claim that this system had "enforced itself on the senior commanders of all armies"[34] is true. But Currie had visited Verdun after the Somme battle of 1916, while Haig had been greatly impressed at a staff conference by the then new French commander, General Robert Nivelle. In any case Haig's consistent fear that a look for himself would show a lack of confidence in his staff planners shows some tunnel vision. He spent October, 1917, when he was deciding to continue his offensive, in high level conferences and political visitations. By November 9, when the need to shore up the Italian front had doomed that plan, except for trying out his tanks at Cambrai, Haig—to give him some credit—was having to field the complaints of his War Minister, Lord Derby, about "the Colonial Forces, . . . especially the Australians, looking upon themselves, not as part and parcel of the English Army but as Allies . . . [like] the Portuguese, not from the point of view of fighting but from . . . [that of] self-control, . . . [with] an implicit belief in [Sir William] Birdwood . . .[as] their Chief in the Field."[35]

Derby's views of an Australian ally who a year earlier had lost 23,000 men in seven weeks at Pozieres, after one division had lost 5,533 men in one night in what GHQ called "some important raids" at Fromelles, and who in the two Anzac corps —had just lost another 38,000 men in five weeks, highlights Eisenhower's sensitivity to national feelings a quarter of a century later. The new Commonwealth of Australia had less than 5,000,000 people in 1914. Terraine's claim that Haig took a new view of Australian discipline in March, 1918—after writing his wife in February that "We have had to separate the Australians into Convalescent Camps of their own, because they were giving so much trouble when along with our men and put such revolutionary ideas into their heads," and complaining of Birdwood's discipline on the basis of a chart which showed that the other colonials had 1.6 men per thousand in prison to the Australians' nine—is highly suspect. Perhaps Haig changed his mind when he used the Australian divisions piecemeal to check the German March offensive, while "the Canadians, despite their undoubted military virtues, insisted on homogeneity, and took no part in the great defensive battle."[36]

A different view of his duties might have led Haig to an earlier appreciation of Currie's artillery tactics, or —with British armies which had rapidly changed from "restricted professional" to "widely conscriptive"—to a different view of Australian "modes of behavior, both in and behind the lines, which never ceased to puzzle him . . . most British officers—and ordinary Tommies, too."[37] As a result of such misunderstandings of the Australian "character" at this particular time, Monash wrote in 1920 that;

> Very much and very stupid comment has been made upon the discipline of the Australian soldier, . . . because the very . . . purpose of discipline [has] been misunderstood . . . . It does not mean . . . obsequious homage to superiors, . . . forms and customs, . . . [but that] unques-

SOME HYPOTHESES

tioning obedience was the only road to successful collective action. He acquired those military qualities because his intelligence taught him that the reasons given him were true ones, . . . [and because] democratic institutions . . . [and an] advanced system of education . . . [had taught] him to think for himself and to apply what he had been taught to practical ends.[38]

How, then, did Haig and Admiral Sir John Jellicoe—who was just as slow to grasp new tactics and technologies—keep the confidence of their British subordinates and of such outsiders as Currie and Monash?[39] The diaries to which Haig confided his frustrations also show that he was a gentleman, who did not berate men in public. Courtesy, Elie Halevy's *History of the English People in the Nineteenth Century*[40] suggests, had become one more cohesive for a changing, but still highly stratified society. That frontier hospitality, another form of courtesy, noted in Alexis de Tocqueville's *Democracy in America*,[41] may have played a similar role in that society. And if Eisenhower was to work out his frustrations with Patton and Montgomery in private Angle-Saxon language, his memoirs were to give only negative evidence that he never really liked John Foster Dulles.

Outside observers of the British Army met a good many regular NCOs and officers whose experience with the least developed elements of British society had not helped them to deal with class-conscious workers. But Birdwood and Haig were not the only British regulars for whom colonials developed a high regard. Many of them had been picked for their prewar jobs by Lord Kitchener, a commander who, on their showing, was a good judge of character. And one often overlooked result of what is often seen as his greatest blunder, his 1914 decision to raise a new army instead of expanding the regulars and territorials, was that most of its lower ranking officers were middle-class gentlemen, like Clement Attlee, who knew the importance of courtesy and face in handling class-conscious workers. As we have already noted, this kind of democratic *noblesse oblige* was probably stronger in the British and French armies than in a German one that was still caught between old traditions and new in securing the "participation of the people in this great affair of state."[42] Monash, a Jew, was impressed by a Lieutenant Colonel Levey, "an active Zionist and very much interested in . . . a Jewish state for Palestine," a member of Sir Ivor Maxse's new training methods team and also *"the officer commanding the Gordon Highlanders!"*[43]

The complexities of the MSC factors for these three historically related but already differentiated armies and societies are obvious. All fought well, though Canada's MPR was lower because of some historic attitudes of French Canada. If their degrees of social stratification were different, a historian has as hard a time as postwar radicals had in finding which had the most class-conscious and most revolutionary-minded workers. All showed great social cohesion, though only Great Britain—less both parts of a still legally united Ireland—adopted conscription, while French Canada showed its cohesion by voting against it for a "British" war. And Australia's soldiers, with 64.8 percent battle casualties to Canada's 49.1 percent and Britain's 47.1 percent of those who "took the field," barely voted for it lest it "dishonor" the volunteer spirit.[44]

Armed Forces and an Aspect of American Life

In the nearly two centuries since this American republic declared its independence, its armed forces have undergone many changes. They began as MsC or "tribal," but the Congress was soon forced to enlist a long service core that proved so useful that George Washington—according to John M. Palmer's *America in Arms*[45]—proposed a similar "restricted professional" nucleus for a better-trained militia. But all his successors got during the next two generations was a very restricted mSC force and a militia which might be graded msC or "polis-type." The professional officers on both sides in the ensuing Civil War created MSC "widely conscriptive" armies, if one agrees that

SOME HYPOTHESES

both nationalisms successfully overcame doubts about the causes for which they were fighting. The postwar generation reverted to tiny professional forces and an even smaller, in proportion to our population, and no better trained militia. The twentieth century's crusades required "widely conscriptive" MSC systems; that of the Second World and early Cold Wars may now be passing from an "impossible" MSc to an equally "impossible" mSc phase, until a new foreign and internal policy consensus brings us back to a mSC "restricted professional" system.

The tensions within the nineteenth century's professional militia mixes are reasonably familiar. Marcus Cunliffe's *Soldiers and Civilians: The Martial Spirit in America*[46] has chapters on "The Confused Heritage and its Continuance," "The Martial Spirit," "The Professionals: Unpopularity" and "Consolidation," "The Amateurs: Apathy and the Militia," and "Enthusiastic Volunteers." But the "three rival yet complementary patterns" of the American ethos worked surprisingly well. "Quaker, Rifleman, and Chevalier correspond to broader divisions of reason, appetite, and sensibility .... To the extent that it neglected the first, the antebellum white South [John Hope Franklin's *Militant South*],[47] enjoyed a greater apparent harmony. This was why some Northern soldiers fancied the Southern atmosphere was more congenial to them."[48]

Militarism was a nineteenth-century Prussian liberal pejorative which Anglo-Americans chiefly applied to Continental Europe's conscript armies. There is no index entry for militarism in the great 1911 *Encyclopaedia Britannica*. John H. Muirhead's article on Hegel ignored Hegel's idea that modern war fosters unselfishness and does not lead individuals to hate individuals, though he did stresss that "overpowering sense of the value of organization" which led Hegel to believe that "a vital interconnection between all parts of the body politic is the source of all good."[49] Great War Anglo-American propagandists saw militarism as characteristically German, but Emory Upton's post-Civil War proposals for compulsory military training and later proposals for a general staff had already been attacked as too Prussian. Cunliffe's "Quakerism" has deep Anglo-American roots, while Tocqueville's fears that the inevitable growth of democracy would lead to despotism and militarism have gradually become linked with twentieth-century fears that any really popular crusade would have the same consequences.

Tocqueville's fears of Bonapartism were inseparable from some of his fears for America. Peace was "peculiarly hurtful to democratic armies," though war and its popular passions gave "them advantages [if they were not defeated immediately] which . . . cannot fail in the end to give them the victory." And "no kind of greatness is more pleasing to the imagination of a democratic people than military greatness—a greatness of vivid and sudden luster obtained without toil by nothing but the risk of life." For these reasons Tocqueville felt that "no protracted war can fail to endanger the freedom of a democratic country," if only because "it must increase the powers of civil government." A democracy's defenses lay in the characteristics of its officers, noncommissioned officers, and men, though these were not uniform "at all times and among all democratic nations . . . . The noncommissioned officers will be the worst representatives of the pacific and orderly spirit, . . . and the private soldier the best." If the "community is ignorant and weak," its soldiers may "be drawn by their leaders into disturbances; . . . if it is enlightened and energetic, the community will itself keep . . . [them] within the bounds of order."[50]

Tocqueville died in 1859, on the eve of the American and German wars and the Triumphs of two kinds of people's armies. During the next long peace the English liberal Hegelian, Thomas H. Green, argued that more democratic states saw the general will less militantly, were less prone to resort to war, and more likely to "arrive at a passionless impartiality in dealing with each other."[51] His Social Darwinist contemporary, Herbert Spencer, held that the progress of democracy and industrialism had already resulted in "a growing personal independence, . . . a smaller faith in governments, and a more qualified patriotism." The general decrease of warfare in the nineteenth century had already resulted in some lessening of governmental controls and national militancy, though Spencer feared feudal and national Germany's increasing "armaments and aggressive activities" and regression "toward the militant social type; alike in the development of the civil organization with its accompanying . . . ideas, and in the spread of socialistic theories."[52]

The Marxist analysis of the relations of armed forces to society, diluted as it was in Anglo-America, was analogous to parts of the Social Darwinism that was part of the theoretical basis for Anglo-American navalism and imperialism. To Karl Liebknecht, *Militarism* "exhibits . . . the national, cultural, and class instinct of self-preservation, that most powerful of all instincts." Military history is that of "human development, . . . strained relations and jealousies between nations and states, . . . [and internal] class struggles for the same objects."[53] While Friedrich Engels' early support of compulsory training to prepare workers for the coming revolution had hardly allayed conservative fears of mass armies' political unreliability, no socialist general strikes prevented the outbreak of war in 1914. That the workers have fought so well when "great interests" have been involved only underlines those populist and nationalist demands for military power that may produce what Morris Janowitz calls "reactive militarism."[54]

If Americans have become "reactive militarists," they have not done so for the reasons suggested in Alfred Vagts' massive 1937 *History of Militarism: Romance and Realities of a Profession*. That work, a German exile's view of a tragic past, separated the scientific "military way" from a "militarism . . . of caste and cult, authority and belief."[55] Huntington's "militarism," twenty years later, was Vagts' "military way." Its American strands were "technism, popularism, and professionalism, . . . [and] the extent to which liberal ideology and conservative Constitution . . . dictate an inverse relation between political power and military professionalism." Huntington's romanticism then credits American professional soldiers with a "military ethnic" that combines a Hobbesian view of "the permanence . . .[of] evil in human nature," a Hegelian "supremacy of society over the individual," medieval ideas of "order, hierarchy, and division of function," and modern ideas of "the nation state as the highest form of political organization, . . . the continuing likelihood of wars among nation states, . . . and the importance of power in international relations."[56] While many American professional soldiers believed the Social Darwinistic or imperialistic slogans of the turn of the century, these were hardly part of an overall military and political philosophy, while equally pragmatic economic arguments against war had an almost equal and just as unphilosophical appeal to such big-business pragmatists as Andrew Carnegie.

Those arguments were best expressed in the Russian Jewish banker Ivan S. Bloch's statistical projections of *The Future of War in Its Technical, Economic, and Political Relations* (1898), a work which helped to persuade Tsar Nicholas II to call the First Hague Peace Conference. The future of war, Bloch predicted, was one of tactical deadlock created by new military technologies, economic collapse, and political and social revolution.[57] As Cunliffe notes for an earlier era, "War is tragic; to the practical Quaker it is also supremely wasteful."[58] In 1910 the American psychologist William James published "The Moral Equivalent of War." While "modern war is so expensive that we feel trade to be a better avenue for plunder, . . . modern man inherits all the innate pugnacity . . . of his ancestors." Could one get "the martial type of character . . . without war," by social service training which would demand the same "strenuous honor and disinterestedness?"[59] Cunliffe notes in his epilogue that American involvement in the Civil War had been "intense yet oddly superficial." In the next generation "American society was possibly even more civilian in outlook than before, . . . [with] captains of industry rather than . . . captains of armies . . . [its] modern heroes."[60] By 1910 the inhabitants of

SOME HYPOTHESES

Disneyland's prosperous family farms and Main Streets were practical Quakers and Riflemen who would still fight, but for a Spencerian ideal world of "peaceful occupations, . . . honesty, truthfulness, . . . [and] kindness."[61] If Disneyland is a romanticized early twentieth-century America, the real America was not evil, cruel, or militaristic.

Huntington's final section sees the "Main Street of Highland Falls" epitomizing a society without "common unity or purpose . . . . West Point embodies the military ideal at its best; Highland Falls the American spirit at its most commonplace. West Point is a gray island in a many colored sea, a bit of Sparta in the midst of Babylon." If Disneyland is romanticized, Highland Falls as Babylon-on-Hudson is a massive misperception. In 1890 there were fewer than 4,000 American officers on active duty—a few more than the 1970 West Point student body—for a population of nearly 63,000,000. In 1938 the first figure had risen to less than 27,000—a quarter of the number of physicians and 6,000 under the 1970 Penn State student body—for a population of nearly 130,000,000. This tiny professional guild was to be so successful precisely because it was so closely attached to a pragmatic, machine-minded idealism. As Huntington had noted earlier, it had had to compensate for its numbers by seeing war "as an independent science, . . . the practice . . . of which was the only purpose of military forces."[62]

The American World War Crusades reflected the militant Calvinism of the frontier Riflemen, Quaker pragmatism, and Cavalier ideas of honor. The American professional soldier did what his society wanted him to do as well as any professional military man in history. Some soldiers were rewarded with the highest offices in the land, but it is a total distortion of history to see Douglas MacArthur's regency in Japan, George C. Marshall's Europèan recovery plan, or Eisenhower's presidency as reflecting a conscious militarism, as Vagts defined that term, or these soldiers as "out of touch with the living, acting, operating nation." And if by 1965—a "peacetime" year for that decade— there were 350,000 American officers on active service for a population of 195,000,000, and if their machines took an historically unprecedented portion of a peacetime Federal budget, all this was at most the "reactive militarism" of a world power which had been pulled into that position by the logic of international politics and pushed in by popular passions and ideals.

Does the historical development of our armed forces "as an important aspect of American society and culture" suggest the possibility of their alienation in the future? My first hypothesis on this point is that we are in greater danger of "reactive pacifism," in a world which still contains people who do not wish us well, than of conscious militarism. The strength of Europe's Fascist movements in the interwar era was that they were primarily civilian. Quakerism is as American as Calvinism. The direct, easily recognizable contributions of armed forces to our society were relatively sporadic during our whole century of "free security" from 1815 to 1914. The Army's contribution to civilizing our barbarians and their lands and the Navy's support of a trade with underdeveloped areas which steadily declined in relative importance had to be rediscovered by such scholars as Francis P. Prucha and Don Rickey, Jr. This may show how remote these contributions had become to Disneyland's second or third generation farmers, or to the emigrants who had poured into our cities. And the appeal of Westerns to their descendants may be more a matter of popular taste than of any deep appreciation of our armed forces' contributions to our society and culture.

My second hypothesis is based on Huntington. Our annual Federal budgetary battles are adversary-advocate charades between the people's representatives and popular demands for public services. On April 15 the ghosts of British taxmen, railroad barons, and Wall Street mug taxpayers on the streets of the District of Columbia. Washington is Fun City when it hands out money to individuals and local political bodies. Indirect job handouts through arms makers or military installations get wide local publicity, but one wonders whether large-scale revenue sharing with every political unit in the land will weaken "reactive" pressures from not necessarily "pacifistic" civilian bodies for other social purposes than defense.

Third, the Crusades of the first half of this century were accompanied by exaggerated fears of plots and heresy. Our armed forces have been largely stationed overseas for more than a quarter of a century. The physical isolation of many home military installations and the armed forces' needs, both real and reactive, for secrecy could fuel a reactive McCarthyism at both ends of the military-civilian spectrum. Stab-in-the-back theories of our Vietnam defeat or victory—in the present climate of opinion almost any noun is pejorative for someone—may be as romantic as *Seven Days in May* scenarios, but such theories and scenarios will not lessen the murkiness of that climate. Fears of a largely black mercenary army are already being exploited. Whether military job and status competition has been a factor in what may appear to be old-fashioned black and white confrontations is one problem of the "Ethno-Cultural Composition of Armed Forces."

My final hypothesis on possible sources of alienation of armed forces from society is even murkier and more general. Does the fact that absolute weapons and assured delivery sufficiently threaten a defender's civilians only slightly less than they do those of the attacker help to explain that mixture of apathy and fear with which civilians seem to view the slide rule computations by which military conflict managers promise civilians security? Cannot both soldiers and civilians complain of the gross but gripping oversimplifications of television, a movable feast which cannot be stopped for chewing on mathematical equations and institutional printwork?

Conclusions

My own model of military history stresses its Political, Military organizational, and Technological variables. Andreski's MSC formula for *Military Organization and Society* deals with the middle term in my model. Since his military Cohesion is so dependent on military Subordination, I would change it to social Cohesion and place it after Politics in a more general formula. By making Technology the first as well as the last term, this becomes a six-element TPCMST closed-loop feedback system for more general conclusions than the usual one that all historical generalization is difficult and dangerous.

Technology is first used in the general sense of the optimal use of food, materials, power, and transport at, say, the hunting, food gathering, Neolithic agricultural, or modern industrial levels for such social purposes as war. Social organization—which is also critical in the history of technology—is also a Political factor. Roman or American settlers, for example, were superior in food production to their barbarian neighbors, while their literate social and Political organizations allowed them to make much better military use of their existing food, materials, power, and transportation resources. And the Romans were also more efficient than other ancient city states or empires because their willingness to expand their citizenship resulted in higher levels of political participation over wider areas.

Andreski's Military Participation Ratio—which can be measured statistically—and military Subordination factors need no further explanation. With Cohesion used in a social rather than in a military sense, one can see how a Political system with low social Cohesion and Military participation and high military Subordination may become unstable to the point of internal revolution when war demands higher Military participation, or how a Technologically declining—in relation to its neighbors—Roman Empire survived by retaining its solid political organization and professional forces' high military subordination and discipline while regaining the social cohesion of the early republican system. This is why Americans can feel that the retention of what they regard as a democratic policy may be critical for military survival. But Technology is, in the short run, politically neutral. Its use can be a powerful factor in the success of a totalitarian system, when such a system concentrates on the use of technology for tactical military purposes. The final T in

SOME HYPOTHESES

my TPCMST formula stresses this point. In an era of technological warfare, absolute weapons, and global delivery systems this last T—the use of advanced Technologies for tactical military purposes—may be as important as the first, or the intermediate P, C, M, and S factors.

Notes

1. Benedetto Croce, *Historical Materialism and the Economics of Karl Marx*, trans. C.M. Meredith (London: H. Latimer, 1914), pp. 3–4.
2. Polybius, *The Histories*, trans. Mortimer Chambers (New York: Washington Square Press, 1966), p. 1
3. Stanislav Andreski, *Military Organization and Society*, 2d ed. (Berkeley: University of California Press, 1968).
4. Karl von Clausewitz, *On War*, trans. O.J. Matthijs Jolles (New York: Modern Library, 1943), p. 18.
5. *Ibid.*, pp. 5–6, 483.
6. Samuel P. Huntington, *The Soldier and the State; the Theory and Politics of Civil-Military Relations* (Cambridge, Mass.: Belknap Press of Harvard University Press, 1957).
7. Andreski, *Military Organization and Society*, pp. 122–123, 244.
8. Chester G. Starr, *The Emergence of Rome as Ruler of the Western World*, 2d ed. (Ithaca: Cornell University Press, 1953), p. 110.
9. Andreski, *Military Organization and Society*, pp. 122, 244.
10. John Monash (Sir), *The Australian Victories in France in 1918*, rev. ed. (London: Hutchinson, 1936), p. 316, first published in 1920.
11. Clausewitz, *On War*, p. 576, Peter Paret, "Education, Politics, and War in the Life of Clausewitz," *Journal of the Histories of Ideas*, 29 (July–September, 1968): 394–408.
12. Polybius, *Histories*, p. 3.
13. *Ibid.*, p. 236.
14. *Ibid.*, p. 232.
15. *Ibid.*, pp. 260–264.
16 Starr, *Emergence of Rome*, pp. 18–21.
17. Andreski, *Military Organization and Society*, pp. 200–201.
18. Stuart Piggott, *Ancient Europe from the Beginnings of Agriculture to Classical Antiquity* (Chicago: Aldine Publishing Co., 1965), p. 256.
19. Arthur E. R. Boak, *Manpower Shortage and the Fall of the Roman Empire* (Ann Arbor: University of Michigan Press, 1955).

20. Archibald R. Lewis, *Naval Power and Trade in the Mediterranean, A.D. 500–1100* (Princeton: Princeton University Press, 1951), p. 25.

21. Walter Goffart, "Zosimus, The First Historian of Rome's Fall," *American Historical Review*, 76 (April 1971): 414.

22. Robert S. Lopez, *The Birth of Europe* (New York: M. Evans, 1966), pp. 19–24.

23. Lynn T. White, *Medieval Technology and Social Change* (Oxford: Clarendon Press), 1962.

24. Alfred T. Mahan, *The Influence of Sea Power upon History, 1660–1783* (Boston: Little, Brown and Co., 1890).

25. John Beeler, *Warfare in England, 1066–1189* (Ithaca: Cornell University Press, 1966), p. 17. Oman's work, published in 1898, was revised and edited by Beeler. Beeler's *Warfare in Feudal Europe, 730–1200* (Ithaca: Cornell University Press, 1971), will be quickly recognized as the standard account.

26. Beeler, *Warfare in England, 1066–1189*, pp. 33–317.

27. Geoffrey C. Evans (Sir), *Tannenberg, 1410: 1914* (Harrisburg: Stackpole Books, 1971), pp. 42–43.

28. Giulio Douhet, *Command of the Air*, trans. Dino Ferrari (London: Faber and Faber, 1943), pp. 125, 128.

29. David L. George, *War Memoirs of David Lloyd George* (London: Nicholson & Watson, 1933-1936), 4: 2100.

30. Hugh M. Urghart, *Arthur Currie: The Biography of a Great Canadian* (Toronto: Macmillan, 1930) will be replaced by the forthcoming study by A.M.J. Hyatt of the University of Western Ontario.

31. Theodore Ropp, *War in the Modern World*, rev. ed. (New York: Collier Books, 1962), p. 222.

32. The English title of Terraine's defense of Haig was *Douglas Haig, the Educated Soldier* (London: Hutchinson, 1963). The American edition is *Ordeal of Victory* (Philadelphia: Lippincott, 1963).

33. George, *War Memoirs*, 4: 2236.

34. Terraine, *Ordeal of Victory*, p. 140.

35. Douglas Haig, *The Private Papers of Douglas Haig, 1914–1919*, ed. Robert Blake (London: Eyre & Spottiswoode, 1952), p. 266.

36. Terraine, *Ordeal of Victory*, p. 218.

37. *Ibid.*, pp. 217–218.

38. Monash, *The Australian Victories*, pp. 314, 315–316.

39. A.M.J. Hyatt, "Sir Arthur Currie at Passchendaele," *Stand-To*, 10, no. 1 (January–February, 1965): 20, shows that both Monash and Currie felt that Haig was wearing the Germans down. For an Australian view of Terraine's defense of Haig see F.C. Green, "Sir Douglas Haig: Another Point of View," *Stand-To*, 9, no. 2 (March–October, 1964): 1–4. Green had risen from private to captain from 1915 to 1918.

40. Elie Halevy, *A History of the English People in the Nineteenth Century*, 2d. rev. ed. (London: E. Benn, 1949–1952).

41. Alexis de Tocqueville, *Democracy in America*, trans. Henry Reeves, 4 vols. (London: Saunders and Otley, 1835–1840).

42. See the later Marshal Lyautey's ideas cited in Ropp, *War in the Modern World*, p. 199.

43. John Monash (Sir), *War Letters of General Monash*, ed. F.M. Cutlack (Sydney: Angus and Robertson, 1935), pp. 272–273.

44. Charles E.W. Bean, *Anzac to Amiens; a Shorter History of the Australian Fighting Services in the First World War* (Canberra: Australian War Memorial, 1946), p. 523.

45. John M. Palmer, *America in Arms: The Experience of the United States with Military Organization* (New Haven: Yale University Press, 1941).

46. Marcus Cunliffe, *Soldiers & Civilians: The Martial Spirit in America, 1775–1865* (Boston: Little, Brown, 1968).

47. John H. Franklin, *The Militant South 1800–1861* (Cambridge Mass.: Belknap Press of Harvard University Press, 1956).

48. Cunliffe, *Soldiers & Civilians*, p. 423.

49. *Encyclopedia Britannica*, 11th ed.

50. Tocqueville, *Democracy in America*, 4: 231–232, 239–240.

51. Thomas H. Green, *Principles of Political Obligation* (London: 1890), para. 175.

52. Herbert Spencer, *The Principles of Sociology* (New York: D. Appleton, 1896–1897), 2, chap. 23.

53. Karl P.A.F. Liebknecht, *Militarism* (New York: B.W. Huebusch, 1917), p. 2.

54. Morris Janowitz, *The New Military: Changing Patterns of Organization* (New York: Russell Sage Foundation, 1964), p. 16.

55. Alfred Vagts, *A History of Militarism: Romance and Realities of a Profession* (New York: Norton, 1937), p. 11.

56. Huntington, *The Soldier and the State*, pp. 79, 143, 193.

SOME HYPOTHESES

57. Ivan S. Block, *The Future of War in its Technical, Economic, and Political Relations: Is War Now Impossible?* (Boston: Little, Brown, 1903). Translated by R.C. Long from volume 6 of the original six volumes.

58. Cunliffe, *Soldiers and Civilians,* p. 414. Block's argument was used most directly in Norman Angell's famous *The Great Illusion: A Study of the Relation of Military Power in Nations to their Economic and Social Advantage* (London: Putnam's, 1911).

59. Use the reprint along with other key works on the subject, in Leon Bramson and George W. Goethals, eds., *War: Studies from Psychology, Anthropology, Sociology* (New York: Basic Books, 1964), pp. 22, 29.

60. Cunliffe, *Soldiers and Civilians,* pp. 435–436.

61. Spencer, *Principles of Sociology*, 2: chap. 23.

62. Huntington, *The Soldier and the State,* pp. 225, 464–465.

COMMENTS

One of the factors that complicates equations between armed forces and societies is technology. In Corelli Barnett's presentation of Jutland as a symbol of the twentieth-century decline of British power, it is the degeneration of British technology —itself a symptom of deeper forces of social decay—that is the proximate cause of the degeneration of the Royal Navy that revealed itself dramatically in combat on May 31, 1916. But technology also has a perverse way of not necessarily declining in direct ratio to a state's social decline; the intractable history of Byzantium again comes to mind, because it was the sick empire's possession of the mysteries of Greek fire that was one of the main sources of its apparently paradoxical military power.

Of course, it has been only in the past few centuries that technological change has played a consistently decisive role in warfare. Daniel R. Beaver here writes about the first phase of the acceleration of technological change that has been so characteristic of the modern era, in the seventeenth century, and he comments upon the impact of that first acceleration of technological change upon modern culture in general and upon warfare as an expression of the culture. Professor Beaver had to prepare his 1971 essay with far less wealth of earlier commentary to draw upon than Professor Ropp had in preparing the preceding essay. The history of technology has only with relative recency begun to thrive as a recognized branch of the historical profession; and until now, with a few exceptions, the history of technology has been remarkably neglected even by military historians. The exceptions include one monument of historical literature, John U. Nef's **War and Human Progress: An Essay on the Rise of Industrial Civilization** *(Cambridge: Harvard University Press, 1950). But military history as usually written deals less with the tools of war than one would expect.*

=== COMMENTS ===

Professor Beaver of the University of Cincinnati is one of the younger of the new military historians. Because he began with the military history of the United States, his present essay offers another illustration of the international nature of the field. He first dealt with war in relation to technology and the whole culture when writing about the first war of industrial mass production, in **Newton D. Baker and the American War Effort, 1917–1919** *(Lincoln: University of Nebraska Press, 1966).*

Cultural Change, Technological Development, and the Conduct of War in the Seventeenth Century*

DANIEL R. BEAVER

"To understand is first of all to unify." —Albert Camus

ONE of the significant issues of our day is the reestablishment of some control over international violence. Indeed, men of the twentieth century who have suffered so much from the impact of totality in war find the problem one that defies resolution. But the existence of violence, and the organized and directed violence embodied in the social institution called war, is a part of the history of civilizations.[1] And history is not a study of steadily escalating violence; there have been moments in time when the levels of violence have de-escalated rapidly. The dynamics of escalation have been well studied.[2] But the processes of de-escalation have not received similar careful attention. Is there some cultural phenomenon that might, at particular times and places, make for diminished levels of war rather than accelerated ones?[3] What explanations, if any, are there for these almost rhythmic fluctuations? One other period, one much like our own, the seventeenth century, offers an opportunity to study both the phenomenon of escalation as it exploded in the Thirty Years War (1618–1648) and the subsequent de-escalation toward limited war in the eighteenth century.[4]

These kinds of questions no longer lend themselves to simplistic explanation. Even a recent introductory text, for example, rejects the notion that the decline in the levels of violence after 1648 was due only to the Thirty Years War.[5] Rather, it seems, the "frightfulness" of war was reduced through the slow reappearance of a set of values that, by emphasizing European commonality, tended to make less significant those ideological and religious differences that for a century and a half had turned Europe into a warring camp and divided nations within themselves. There was a reestablishment of community in Western Europe, and men of that day realized it had occurred.[6] During the sixteenth-century religious upheavals in France, Blaise de Montluc, a marshal of the king, wrote: "Toward an enemy all advantages are good, and for my part (God forgive me) if I could call all the devils in Hell to beat out the brains of an enemy that would beat out mine, I would do it with all my heart."[7] Thomas Hobbes, writing in the midst of the English Civil War, analyzed in a classic statement the spiritual effect of a generation of violence:

> . . . the laws of nature, as justice, equity, modesty, mercy, and, in sum, doing to others what we would be done to, of themselves, without the terror of some power to cause them to be observed, are contrary to our natural passions, that carry us to partiality, pride, revenge, and the like. And covenants, without the sword are but words, and of no strength to secure a man at all.[8]

But little more than a hundred years later, toward the end—unknown to him, of course—of a period of relative stability in Europe, the English historian Edward Gibbon would write:

CONDUCT OF WAR IN THE SEVENTEENTH CENTURY

> ... a Philosopher may be permitted to enlarge his views, and to consider Europe as one great republic whose various inhabitants have attained about the same level of politeness and cultivation. The balance of power will continue to fluctuate, and the prosperity of our own or the neighboring kingdoms may be alternately exulted or depressed, but these partial events cannot essentially injure our general state of happiness, the system of arts, and laws, and manners, which so advantageously distinguish, above the rest of mankind, the Europeans and their colonies .... In peace, the progress of knowledge and industry is accelerated by the emulation of so many active rivals; in war, the European forces are exercised by temperate and indecisive contests.[9]

Just as Hobbes described a particular moment in time as a universal condition, so Gibbon, on the eve of the French Revolution, looked back on the previous one hundred years in the belief that he was looking forward as well. He could not know that he was describing a unique period when the forces of community had created a temporary equilibrium in Western European culture that would not appear again for several generations.

The story of how this equilibrium of community appeared, how the Heavenly City of Latin Christendom was melded over two centuries into what Carl Becker called the *Heavenly City of the Eighteenth Century Philosophers*, concerns the relationship of technology, internal military organization, and ideas and their incorporation ultimately into administrative systems that held together until the catastrophic shock of the French Revolution. By the beginning of the seventeenth century the technology necessary for an arms industry was available; the lathe, the boring machine, the trip-hammer, the die, and the jig had all been invented. Although a satisfactory prime mover had not yet been developed, the waterwheel, the windmill, and the capstan were available to supplement human muscle and horsepower. Just as gunpowder and the missile weapon extended human reach, wind and water extended human strength. The science of metallurgy was as yet imperfectly understood, but sufficient was known to produce metal—both bronze and iron—for reliable small arms and artillery.[10]

The introduction of missile weapons and the adoption of gunpowder offered a singular opportunity for the emerging dynastic states to gain a monopoly over their manufacture. Royal monopolization of arms production, especially artillery, when linked with the appearance of standing armies goes far toward explaining the process of dynastic aggrandizement so characteristic of the seventeenth century. Such attempts were not new; during the seventeenth century they simply became more effective.[11] In France, under Henry IV's great principal minister, Maximilien de Béthune, duc de Sully, attempts to root out illicit manufacture of gunpowder, saltpeter, and artillery pieces were largely successful. The Ordinance of 1601 declared that the right to make saltpeter and gunpowder was as much an attribute of sovereignty as the right to coin money. The management of the royal saltpeter and gunpowder manufactures was placed in the hands of the Grand Master of Artillery, an office that Sully himself held from 1599 to 1610.[12] Despite some increases in private illegal production of gunpowder during the *Fronde,* royal control never collapsed completely. In 1663 Louis XIV issued an ordinance insisting upon the enforcement of previous laws, and under the improved administration of his reign the Grand Master of Artillery was in an excellent position to control the production of munitions.

In England gunpowder manufacturing did not become critically important until Elizabeth's reign. In the 1560s the sole production of gunpowder was invested by monopoly patent in George Evelyn. Until 1635 the monopoly remained in the hands of his wealthy Middlesex descendants. During the reign of Charles I the monopoly began to break down, until in 1641 Parliament threw the manufacturing of gunpowder open to the entire community.[13] Thus, by the end of the seventeenth century control over munitions in England had taken an entirely different course from that in France, where its manufacture was under royal supervision from beginning to end. The English government contracted on the private economy, and the English kings thereby failed to keep gunpowder production monopolized in their own hands. The administrative machinery was systematized sufficiently in England to assure adequate munitions to the king's armies, but through cooperation with private entrepreneurs rather than through the outright monopolization of manufacturing.

There were substantial improvements in weapons before 1675. The flintlock musket replaced the matchlock; a rational system of light, field, and siege artillery appeared; calibers were to a certain degree standardized; self-contained cartridges for both small arms and cannon were introduced; and, finally, the ring bayonet was developed. But the actual manufacturing of small arms and artillery was a slow, laborious process. Accumulation was through stockpiling rather than through rapid production. After the design of the flintlock musket with either the plug or ring bayonet was stabilized, there was relatively little change in small arms for over a century. The soldier carried the same simple iron tube with a relatively slow rate of fire in 1790 that he had carried in 1690. Artillery reached a similar technological plateau.[14] As a result, so long as reasonable care was taken, arms stored in royal arsenals for decades could be issued to troops without affecting battlefield performance. When the French Army was reequipped with the ring bayonet in the early 1690s, the old muskets that were not machined to the recent innovation were stored. In 1703 they were sold to the Spanish government in large numbers together with their plug bayonets and were used effectively by the Spanish army from 1703 to 1710.[15]

It was in the organization and employment of infantry that the relationship of technology and tactics was best articulated. The proportion of pike to shot was decreased and the proportion of artillery to troops was increased. The Spanish *Tercio,* a clumsy phalanx, was modified by the Swedes into a smaller, more maneuverable unit, the battalion and, ultimately, by the

French into the regiment. Volley firing, which maximized the potential firepower of the smoothbore musket, required considerably more training than the old system in which little was required of the soldier except to lean in the proper direction during the "push of pike." At least three years, it was believed, were required to train an infantryman in the new "linear tactics": the complicated process of moving from column into line of battle, from line of battle into square for protection against cavalry, and back again. The discipline to exchange volleys at less than seventy-five paces as at Blenheim (1704) or Fontenoy (1745) was not imposed in a few months. Thus, soldiers became less expendable than during the Thirty Years War.[16]

The technological and organizational capability to impose damage escalated substantially during the seventeenth century. Armies were larger in 1700 than they had been in 1630. But intent de-escalated. I have already suggested that the expense of training and equipping an army played some part. The nature of the revitalized officer corps was also significant. The new standing army provided a reason for being for the old nobility. Their military effectiveness as heavy cavalry destroyed by gunpowder, their political power hemmed in at least by the expansion of the state, many found a place as officers in the king's service. Cadet schools, where the new tactics were taught, were in existence in both France and Prussia before the turn of the eighteenth century. Annual maneuvers revealed how effectively the lessons had been learned. Throughout Western Europe the same tactics and, incidentally, the same values were inculcated into the young aristocrats. A European brotherhood-at-arms had emerged. Chivalry, in another, secular, guise had reappeared.

The "new warfare" might be symbolized by the seigecraft of Sevastian le Prestre de Vauban. As predictable as the mating dance of the praying mantis, as stylized as a problem in geometry, Vauban's siegecraft assured success while reducing casualties for all involved. What one historian has described as the "Toils of Euclid" seemed so inexorable that Louis XIV, in the interest of military economy, changed the rules of siege to allow a commander to surrender honorably after a small breach had been made in a fortress wall and one assault repulsed. War had become "scientific," governed by a set of rules and a system apparently as precise as the geometric approach trenches of a Vauban siege.[17]

The importance of the ideas of "law" and "system" to the man of the seventeenth century was at the heart of the matter. The realm of nature became more fully separated from theology as a field of human inquiry. Francis Bacon embodied the new empiricist spirit when he wrote in *Novum Organum:*

> For the matter at hand is no mere felicity of speculation, but the real business and fortunes of the human race, and all power of operation. For man is but the servant and interpreter of nature. What he does and what he knows is only what he has observed of nature's order in fact or in thought; beyond this he knows nothing and can do nothing. For the chain of causes cannot by any force be loosed or broken, nor can nature be commanded except by being obeyed. And so those two twin objects human knowledge and human power really meet in one; and it is from ignorance of causes that operation fails . . . .[18]

Descartes in his *Discourse on Method* reiterated Bacon's concern about the practical and pragmatic impact of scientific investigation:

> Although my speculations greatly pleased me, I believe that others also had speculations, which perhaps pleased them more. But as soon as I had acquired some general notions concerning physics and when, beginning to test them in diverse difficult particulars, I remarked whither they might lead and how they differed from the principles in present use, I believed I could not keep them so concealed without greatly sinning against the law which obliges us to procure as much as lieth in us the general good of all men, for they have shown me that it is possible to arrive at knowledge which is very useful in life, and that instead of the

speculative philosophy which is taught in the schools, a practical philosophy may be found, by means of which, knowing the power and the action of fire, water, air, stars, heavens, and all other bodies which environ us, as distinctly as we know the various trades and crafts of our artisans, we might in the same way be able to put them to all the uses to which they are proper, and thus make ourselves, as it were, masters and possessors of nature.[19]

From this spirit of inquiry so characteristic of men of that age came a belief that men could develop technical means to control nature, a new attitude that would ultimately result in intellectual self-confidence and a sense of community unknown since the breakdown of medieval Latin Christendom. To see the world as understandable through human reason rather than faith; to see it as held by immutable laws that could be discovered and, if not manipulated, at least conformed to, brought a wave of confidence into the knowledgeable ranks of a society splintered by religious conflict, shriven by widespread warfare and indiscriminate destruction.

The idea of universal laws of nature seemed reflected in the growing unity of dynastic states. By mid-century, Europe seemed to be sorting itself out according to some only slightly understood internal law. Even during the most violent religious conflicts of the previous century, "system" had competed with the chaos associated with the breakdown of medieval community. Under the Tudors in England the local power of the barons had been limited and a pattern of central administration had emerged, particularly in finance and naval affairs.[20] The Swedes had avoided the worst effects of local feudal autonomy; the army that Gustavus Adolphus led into Germany was raised by conscription and at first supplied through a centralized logistical service.[21]

The Louvre in Paris, where the French War Office was located, symbolized, as much as that great baroque pile, Versailles, the new outlook. The Marquis de Louvois was adviser of the crown on military affairs as also the chief administrator of the armed forces. As Jean Batiste Colbert organized and encouraged the industrial, commercial, and financial communities of France, so the Minister of War organized its military affairs. At Louvois' death in 1691, the French War Office included five bureaus each headed by an assistant to the secretary as well as a substantial archives.[22]

In England a similar process occurred. The first "Military Secretary" was appointed by Charles II while he was in exile. When the monarchy was restored, that office became for a time simply the Military Secretary to the Commander in Chief of the Army. During the reign of Queen Anne (1702–14) the name of the office was changed to Secretary of War and its authority linked directly to the Crown. Under the able administrator Sir Robert Walpole, it became the "channel of reference on all questions between the civil and military parts of the country . . . and . . . the constitutional check interposed for regulating their intercourse."[23]

Among the best examples of the move to limit the impact of war was the appearance in European armies of formal Articles of War. By 1700 these articles differentiated sharply between combatants and noncombatants, described the proper treatment of officer prisoners, and prescribed punishment for plundering and pillaging.[24] By the end of the seventeenth century permanent, well-regulated armed forces were at the command of the major European sovereigns. Governments constructed and supplied standardized arms, rations, clothing, tents, and field bakeries to their troops. Central administrative control, although inefficient, inept, and certainly corrupt by modern standards, was adequate to make armies in the field effective.

The center stone that tied together the arch of ideas and administration was the conception of a "Law of Nations" that governed the relationships of sovereigns. Throughout the seventeenth century jurists struggled with the problem of an international order without a common religious belief. The search culminated in what the eminent eighteenth-century lawyer, Emmerich de Vattel, called "the voluntary law of nations." Denying the existence of a sovereignty beyond the state, he wrote: "Because there is no judge, recourse must be made to rules whereby warfare may be regulated." The idea assumed a commonality of interest which tended to de-escalate violence. Moderation was the keynote:

> A treaty of peace can be no more than a compromise. Were the rules of strict and rigid justice to be observed in it, so that each party should precisely receive everything to which he has a just title, it would be impossible ever to make a peace .... Since ... it would be dreadful to perpetuate the war, or to pursue it to the utter ruin of one of the parties,—and since however just the cause in which we are engaged, we must at length turn our thought towards the restoration of peace, and ought to direct all measures to the attainment of that salutary object,—no other expedient remains than that of coming to a compromise respecting all claims and grievances on both sides, and putting an end to all disputes, by a convention as fair and equitable as circumstances will admit of. In such convention no decision is pronounced on the original cause of the war, or on those controversies to which the various acts of hostility might give rise; nor is either of the parties condemned as unjust,—a condemnation of which few princes would submit;—but, a simple agreement is formed, which determines what equivalent each party shall receive in extinction of all his pretensions.[25]

In 1956, Michael Roberts of Queens University, Belfast, published a remarkable essay on the relationship of technology and tactics which he entitled "The Military Revolution 1560–1660." In it he represented the spirit of the age as being embodied in the strict discipline and elaborate mechanical drill required by the new linear tactics, which matched the tendency of the times toward absolute government. "It was certainly no accident," he wrote, that Louis XIII should be "passionately fond" of drill; nor was it a mere personal quirk that led Louis XIV to cause a medal to be struck, of which the reverse displayed him in the act of taking a parade, and correcting with a sharp poke of his cane the imperfect dressing of a feckless private in the rear ranks. The newly acquired symmetry and order of the parade ground provided for Louis XIV and his contemporaries the model to which life and art must alike conform. And the *pas cadence* of Martinet's regiments echoed again in the majestic monotony of interminable alexandrines.[26]

Military life, like the emergence of absolute monarchy was only a part of the process of systematization characteristic of the age. Newton's *Principia,* published in 1687, described a pattern of order regulated by natural law. Newton rejected even divine intervention in his system. He almost jeered at more orthodox contemporaries who believed "The Machine of God's making is so imperfect . . . that he is obliged to clean it now and then by extraordinary intervention and even to mend it as a clockmaker mends his work."[27]

Newton's "system" was part of a cultural phenomenon that reestablished a conception of community and a system of values shared by the leaders throughout Europe that tended to temper the impact of war. Geography, rather than spiritual value, was the issue. Out of this shared value system emerged, in Clausewitz's terms, "limited war" rather than "absolute war." The new unity was caught best by Alexander Pope who, in 1730, published a couplet entitled "Isaacus Newtonus."

Nature and nature's laws lay
hid in night:
God said, Let Newton *be!*
And all was light.[28]

The idea of community, like most of the great ideas that motivate men, defies precise definition. If community exists, men tend to consider it eternal; if it doesn't, its existence seems impossible to conceive. Men of the mid-eighteenth century were part of a buoyant, self-confident age that they could not really believe would end. Neither Newton, nor Hobbes, nor Gibbon, nor even Alexander Pope could see around the corner into the future as History. None could foresee the destruction of community that would come with the revolution in France. Like all of us, they tried to fix a moment and label it eternal. But history is dynamic. It does move. A most tolerant man of that day, Blaise Pascal, described their dilemma and ours:

> Nothing stands still for us. This is the state which is ours by nature, yet to which we least incline: we burn to find solid ground, a final steady base on which to build a tower that rises to Infinity; but the whole foundation cracks beneath us and the earth splits open down to the abyss.[29]

CONDUCT OF WAR IN THE SEVENTEENTH CENTURY

Notes

*The broad conception for this essay was suggested by the studies of two of the most gifted historians of our day, G.N. Clark, *War and Society in the Seventeenth Century* (Cambridge, England: Cambridge University Press, 1958) and John U. Nef, *War and Human Progress* (Cambridge: Harvard University Press, 1950). Their work has raised basic issues about warfare and culture that demands further analysis by anyone concerned about the historic role of violence in societies.

1. The conception of war as a civilized institution is drawn from Quincy Wright, *A Study of War*, 2 vols. (Chicago: University of Chicago Press, 1972).

2. The best example of the study of the "escalator effect" lies in the literature of the First World War. See, for example, Fritz Fischer, *Germany's Aims in the First World War* (New York: Norton, 1967), and Raymond Aron, *The Century of Total War* (Garden City: Doubleday, 1954). For a provocative view of the social dynamics involved in the outbreak of the First World War, see Arno J. Mayer, *Dynamics of Counterrevolution in Europe, 1870–1956: An Analytic Framework* (New York: Harper and Row, 1971), pp. 134–149, particularly.

3. By culture I mean something far beyond common language and geographical proximity. I mean a set of values such as those described by Pitrim A. Sorokin in *The Sociology of Revolution* (Philadelphia: Lippincott, 1925), which form the intellectual and moral mortar that bind societies to each other inside a common culture.

4. By limited war I mean war limited in its objectives and conducted by particular, designated elements of a society against particular, designated opponents. The concept has little to do with casualties. There is little reason to believe that the rate of casualties in combat was lower during the War of the Spanish Succession (1702–1713) than during the Thirty Years War, but the number of noncombatant casualties dropped dramatically.

5. See David Maland, *Europe in the Seventeenth Century* (New York: St. Martin's, 1966).

6. I have accepted the description of community in Erich Kahler, *The Tower and the Abyss, An Inquiry into the Transformation of Man* (New York: Viking, 1967), p. 7. "Communities form traditions; they are the soil, the constantly shifting ground in which the individual is rooted. Indeed everyone's psyche reaches down to the very origin of life; it is at any moment as deep as life itself."

7. Viscount Montgomery of Alamein, *A History of Warfare* (Cleveland: World, 1968), p. 552.

8. Thomas Hobbes, "Leviathan," in Saxe Commins and Robert N. Lonscott, eds., *Man and the State: The Political Philosophers* (New York: Random House, 1947), pp. 3-4.

9. J.F.C. Fuller, *The Conduct of War 1789-1961* (New Brunswick: Rutgers University Press, 1961), p. 26.

10. Lewis Mumford, *Technics and Civilization* (New York: Harcourt, Brace, 1934), p. 88.

11. John U. Nef, *Industry and Government in France and England, 1540-1640* (Philadelphia: American Philosophical Society, 1940), pp. 58-60.

12. Nef, *Industry and Government*, pp. 60-62.

13. Nef, *Industry and Government*, pp. 96.

14. For a description of technology and tactics in the seventeenth century, see Theodore Ropp, *War in the Modern World* rev. ed. (New York: Collier Books, 1962), pp. 19-59. See also Lynn Montross, *War Through the Ages* third ed. (New York: Harper, 1960), pp. 262-390.

15. Henry Kamen, *The War of Succession in Spain 1700-15* (Bloomington: Indiana University Press, 1969), pp. 57-80.

16. John B. Wolf, *Louis XIV* (New York: Norton, 1968), pp. 205, 482. For example, Louis XIV voiced concern for the lives of his soldiers on many occasions in private as well as in public, although admittedly he was as worried about money as about lives.

17. Ropp, *War in the Modern World*, p. 42. For a discussion of Vauban and seventeenth-century France, see Henry Guerlac, "Vauban: The Impact of Science on War," in Edward M. Earle, ed., *Makers of Modern Strategy: Military Thought from Machiavelli to Hitler:* (Princeton: Princeton University Press, 1943), pp. 26-48.

18. Frederick Klemm, *A History of Western Technology,* tr. Dorothea Waley Singer second ed. (New York: Scribner, 1959), pp. 173-174.

19. Klemm, *A History of Western Technology*, pp. 179–180.

20. For a discussion of English naval and financial affairs, see J.A. Williamson, *The Age of Drake* (New York: Barnes & Noble, 1960), Garrett Mattingly, *The Armada* (Boston: Houghton Mifflin, 1959), and Conyers Read, *Lord Burley and Queen Elizabeth* (New York: Knopf, 1960).

21. The best study of Gustavus Adolphus is Michael Roberts, *Gustavus Adolphus: A History of Sweden, 1611–1632* 2 vols. (New York: Longmans, Green, 1953–1958). See also C.V. Wedgwood, *The Thirty Years War* (London: Jonathan Cape, 1938).

22. Guerlac, "Vauban," p. 28.

23. Major R.E. Scouller, *The Armies of Queen Anne* (Oxford: Clarendon Press, 1966), p. 20.

24. For an example of contemporary articles of war, see *Ibid.*, p. 388.

25. Fuller, *Conduct of War*, pp. 18–19.

26. Michael Roberts, "The Military Revolution, 1560–1660," in David B. Ralston, ed., *Soldiers and States, Civil-Military Relations in Modern Europe* (Boston: Heath, 1966), p. 18.

27. A. Wolf, *A History of Science, Technology and Philosophy in the Sixteenth and Seventeenth Centuries* (New York: MacMillan, 1935), pp. 674–675.

28. Howard Lowry and William Thorpe, eds., *An Anthology of English Poetry* 2nd ed. (New York: Oxford University Press, 1956), p. 442.

29. Quoted in Kahler, *Tower and Abyss*, unnumbered page of Preface.

III

National Security and Military Command

French National Security
Policies, 1871–1939
RICHARD D. CHALLENER

page 92

The Japanese Army Experience
ALVIN D. COOX

page 123

High Command Problems
in the American Civil War
WARREN W. HASSLER, JR.

page 152

American Command and
Commanders in World War I
EDWARD M. COFFMAN

page 176

COMMENTS

From sweeping considerations of the whole range of problems arising between the soldier and his society, we turn to more specific problems of civilian control and of military command, and from surveys of many times and places to studies of particular nation-states.

For obvious reasons, much of the post-World War II new military history first concerned the relationship between the soldier and the state in Germany, the birthplace of that modern domination of the state by the soldier that is called militarism. Even before the world's second twentieth-century confrontation with the fearsome combination of an efficient and aggressive German military system linked to an aggressive German state, Germany's own masterful forerunner of the new military historians, Gerhard Ritter, had begun the magisterial study of German militarism that he went on writing amid the trials of war and concluded amid the dislocations of postwar Germany: **Staatskunst und Kriegshandwerk: Das Problem des "Militarismus" in Deutschland** *(4 vols., Munich: R. Oldenbourg, 1954–1968); translated by Heinz Norden as* **The Sword and the Sceptre: The Problem of Militarism in Germany** *(4 vols., Coral Gables: University of Miami Press, 1970–1972). After the Second World War, the problem of German militarism was examined by eminent historians of other nations, especially notably by the British historian John W. Wheeler-Bennett in* **The Nemesis of Power: The German Army in Politics, 1918–1945** *(New York: St. Martin's, 1954), and the American Gordon A. Craig in* **The Politics of the Prussian Army, 1640–1945** *(New York: Oxford University Press, 1956). It is probably in part an indication of how thoroughly the German military problem has already been examined by these and other works that none of the historians visiting the New Dimensions course addressed themselves directly to Germany.*

But France has played as central a part as Germany in

shaping the modern issues of the control of military power both within the state and when unleashed upon the outside world. The French Revolution generated the concept of the nation in arms and created the democratic mass army. The mass army in turn began carrying warfare beyond its narrow eighteenth-century limits, as a contest between small forces of professional soldiers little touching civilians, toward its modern totality. The Prussians then yoked the power of the mass army to the efficient direction of a modern professional officer corps, whose professionalism crystallized particularly in the corporate intelligence of the General Staff. In that process, however, the Prussian General Staff established an ascendancy not only over the Prussian army, but over the Prussian state as well, as a state within the state, on which the civil state was so acutely dependent that in a crisis the autonomous military state wielded more power than the civil state to which it was ostensibly subordinate. This Prussian combination of forces so resoundingly defeated France in the War of 1870–1871, however, that subsequently all the continental powers of Europe felt obliged for their self-protection to adopt some approximation of the Prussian combination of the mass army and the directing corporate intelligence of the general staff—a combination now rendered more terrible in its homeland by the magnification of Prussia into the German Empire.

France especially felt moved to emulate the Prussian military arrangements, in her special peril as Germany's ancient rival for the domination of European politics, and both to fend off a repetition of the 1870–1871 humiliation and in the hope of one day gaining revenge. But in Prussia, the ideology of the eighteenth-century democratic revolution had never firmly rooted itself; there the mass army adopted after the French Revolution was democratic only in its massiveness, not in ideology or principles of control. In France, in contrast, the *eighteenth-century democratic revolution had left a deep and permanent impress upon the national character, even if its ideology was not unanimously accepted. Consequently, in France the effort to graft Prussian military institutions upon a democractic Republic created tensions that jeopardized both the inner stability of the state and the very security against outward foes that the borrowings from Prussia were designed to assure. The history of those tensions is in part uniquely French, but in part also it is representative of the problems posed by Prussian-style militarism for all the great powers. In 1971 that history was succinctly reviewed and commented upon by Richard D. Challener of Princeton University, still another exemplar of the international outlook of the new military historians, the author both of* Admirals, Generals, and American Foreign Policy, 1898–1918 *(Princeton: Princeton University Press, 1973), and of* The French Theory of the Nation in Arms, 1866–1939 *(New York: Russell, 1955).*

French National Security Policies, 1871–1939

RICHARD D. CHALLENER

WHENEVER you start to discuss national security policy, you must immediately move beyond purely military considerations. In fact, the most important job of anyone who lectures on this subject is to demonstrate how a whole host of political, economic, and social factors enter into the determination of the national security policy of any nation. This is a special cause of mine, for I really think that military history becomes a truly vital subject only when it moves well beyond purely strategic or tactical studies to include the full range of nonmilitary factors that affect it.

My topic is the evolution of French national security policies from, roughly, the middle of the nineteenth century to the outbreak of World War II. In one respect, it's an easy topic, for French national security has one single, dominant theme: namely, the defense of France against her so-called "natural enemy," Germany. This provides a focus that is lacking in many other countries. For example, there isn't the same tension that you find in American national security policy between those who stressed Far Eastern interests and those who emphasized European concerns. Nor do you encounter as many differences between the outlooks of the Army and the Navy as you find in the United States. In our country the Navy, since the beginning of the twentieth century, has always had a more Far Eastern orientation than the Army. But in France the Navy was largely an adjunct of an army-based, continentally oriented national security policy intended to defend the nation from Germany.

Let me emphasize the fact that French national security policy was continentally oriented. It is of course true that the French, like all of the other European nations, did try to build an overseas empire in the latter decades of the nineteenth century. There is a vast literature on the subject of European imperialism and the rivalries that grew out of European expansion into Africa and Asia. The French were involved in that movement, and there were, indeed, a number of enthusiastic French expansionists—Jules Ferry comes immediately to mind—who really believed in the cause of empire. But the number of men who shared Ferry's enthusiasm was strictly limited, and the Chamber of Deputies was quick to call them to order whenever it looked as if their overseas ventures might imperil French security interests on the European continent. When the chips were down—when, for example, the Fashoda incident threatened a British-French clash over the upper Nile—the French, more than other Europeans, were prepared to sacrifice a colonial interest if it seemed to threaten their security in Europe. Similarly, there was a strong tendency in France to look at the colonies in terms of the help they could provide in strengthening continental security. For example, when the French dealt with the question of conscripting African natives into their armed forces, they were most concerned with devising ways whereby black troops could be used in a European war fought against continental enemies.

This continental tradition was strong and persistent and, moreover, had long-range implications. In 1940, with the German armies pouring toward Paris, French military leaders were still so wedded to thinking about security and war in continental terms that the idea of continuing to fight from North Africa was too strange, too foreign, too unconventional to command much attention. The idea of going off to North Africa was unthinkable. In fact, there really isn't any significant break with this outlook until the post-World War II era. That is, it was only with French involvements in Indochina and Algeria that a genuine colonial perspective began to emerge, only then that the colonial world began to impose upon the Metropole.

The basic outline of French national security policy is remarkably consistent from the 1860s until the outbreak of World War II. The most logical point at which to begin is the time of the Austro-Prussian War of 1866 when the French Emperor, Napoleon III, discovered—and discovered to his great dismay—that France was losing her dominant position on the European continent. At the beginning of that conflict Napoleon still believed that, regardless of the outcome, France would hold the balance of power in Europe. But when Bismarck's Prussia had overhelmed the Austrians in a mere six weeks, even Napoleon had to realize that France was no longer the foremost European military power.

This is the true starting point for the problem of modern French security. Heretofore, the French had been the people who were able to export their military power as well as their political and ideological influence throughout Europe—as witness the French position at the time of Louis XIV or of the first Napoleon. But now the situation was changing rapidly and drastically. On France's eastern frontier there was emerging a unified Germany that not only possessed a formidable military establishment but was also blessed with a larger population and greater industrial potential.

The question, then, for France was how to adjust to this new and dangerous situation. What kind of national security policy should be adopted to meet the new international and military situation that confronted her? Napoleon III, to be sure, never really came to grips with the issue or achieved any practical solutions. He made a few ineffective military reforms, misplayed his diplomatic cards, and listened to some remarkably bad advice from his generals. In 1870 his armies were crushed almost as readily as the Austrians, his regime collapsed, and, eventually, the Third Republic emerged shakily from the ashes of the Second Empire.

This set the stage for a series of major military reforms that were carried out in the 1870s by the new Republic. These reforms reflected an awareness that Germany was now not only a unified country but also a nation with industrial resources France did not possess and a larger population base from which to draw an army. Nevertheless, the French leaders, military as well as civilian, who enacted these reforms tended to believe that the problem of national security was clear, relatively easy to solve, and manageable primarily in military terms.

The basic military reform, the legislation of 1872 that established a system of compulsory service, was itself the product of three principal factors. First of all, it was believed by virtually all Frenchmen that the principal "lesson" of 1870–1871 was that France had not organized her manpower effectively. All political parties, whether conservative or liberal, agreed that the professional army had failed and that, therefore, the one essential reform was to provide a mass army through the instrument of compulsory military service. In the interest of national security, all young Frenchmen must be obligated to serve. This also implied the necessity to model the new French Army at least in part after the successful Prussian example, for it was agreed that one of the principal reasons for German success in 1870 was the Prussian use of conscripts. Secondly, with the advent of the Third Republic, there was a revival of the tradition of the citizen-soldier, of the concept of the nation in arms. The French, to be sure, had "invented" the concept during the French Revolution, and, after 1792 the conscript armies of the revolution had spread French power throughout Europe. But by 1815 the French people had had enough of war and Napoleonic glory. The bourgeoisie who dominated the post-1815 regimes wanted to keep their sons out of the army and had opted for a professional military force. But after 1871 the revolutionary tradition was revived; there was not only the failure of the professionals to point to, but also the example of the resistance that Léon Gambetta's "republican army" had put up after the defeat of the regulars. Third, and no less important, French military theorists began to argue that henceforth all warfare was going to be mass war fought by mass armies. The French, who had long revered Jomini, now discovered Clausewitz and interpreted him to mean that Europe had emerged into an era of mass war.

Overall, the 1870s were a period of military regeneration. The Army enjoyed a prestige it hadn't known in years—as evidenced by the way in which the sons of the nobility now came back into the officer corps. Nor did military reform stop with the establishment of conscription. It was also believed that the defeat of 1870 could be attributed to failings in the staff system, to shortcomings in the training of officers, and, especially, to a failure to organize French railroads for the purposes of war. So, in the 1870s, military schools were remodeled, a staff system was created, and an entire section of that staff was put to work drafting detailed mobilization plans that would fully exploit the mobility provided by the French rail network.

There was, to be sure, nothing truly original in the French solution to the problem of national security in the post-1870 era. So powerful was the Prussian example that virtually all the European nations, except, of course, Great Britain, built national security systems based upon conscription and the mass army. All Europe did move into an era in which each nation believed that its security rested upon having the maximum number of men under arms at the time of mobilization. Each of the great powers also developed intricate mobilization plans that maximized the speed with which the reserves could be called up and the mass army deployed to the field of battle. All plans, moreover, were founded on the prevailing assumption that wars would be decided—and speedily decided—in great, pitched battles between these mobilized conscript masses. And, it should be emphasized, national security policy was almost exclusively "military." That is, neither military theorists nor general staffs paid much attention to the economic, social, or even political aspects of war: national security was defined almost solely in terms of vast armies, detailed mobilization plans, victory or defeat on the field of battle.

NATIONAL SECURITY AND MILITARY COMMAND

Yet the process of building the French national security system after 1870 was not quite as easy or simple as it may have appeared. In the first place there was always a constant political struggle between those who wanted a more democratic form of conscription and those who resisted such liberalizing tendencies. French conservatives might approve conscription, but they wanted a five-year term of service—which meant many special privileges, minimal tours of duty, and even exemptions for the sons of the wealthy. They argued, as a matter of military principle, that the citizen-soldier and the reservist called back to active duty would not be reliable in combat. They might be adequate to carry out a policy of passive military defense, but they lacked a true offensive spirit—élan vital, conservatives called it—and were militarily inferior to soldiers who had been under arms for a long period of time. Republicans and liberals insisted, with no less vehemence, that the period of military service must be of shorter duration, equal for all, and with few if any exceptions. They argued that, as the state itself became democratized, so too must French military institutions be democratized. The Army, in short, must be a mirror of republican society, and the principle of equality must be faithfully observed. In liberal opinion, the citizen-soldier possessed military virtues that were lacking in the professional or the five-year veteran.

Behind this continuing debate there was always a political motive. Conservatives prized the long-service conscript because they felt that he would be a more reliable instrument of a nationalist, anti-German foreign policy. Moreover, indoctrinated and disciplined by his years of training, he would readily obey orders if commanded to intervene in domestic affairs to put down a strike or quell a political disorder. Liberals, on the other hand, believed that an army of citizen-soldiers, an army of short-term conscripts, was the best guarantee against military praetorianism. Fearing the professional soldier and the presumed tendency of the officer corps to ally with the political Right, French republicans were convinced that a democratic army was necessary to prevent military intervention in domestic affairs. Anyone who has the patience to go through the almost endless pages of debate in the Chamber of Deputies over the conscription legislation of 1872, 1889, 1905, and 1913 can readily be pardoned if at times he comes to think that these political questions were the only issues ever discussed and that the problem of military effectiveness was, at best, a secondary consideration.

In the late 1890s and early 1900s the Dreyfus case seemed confirmation to the republicans and their allies that the military were in league with ultra-Catholics and the Right to subvert the Republic itself. When the supporters of Captain Alfred Dreyfus finally won out, they embarked at once on a militant program to democratize the military services, reduce the length of service from three to two years, and, above all, put a leash upon officers suspected of anti-republican leanings. It can in fact be argued that the victorious republicans pushed their program with such zeal that army morale was adversely affected and French military security for some time weakened.

There is, to be sure, a vast literature on the Dreyfus era which paints a dramatic picture of a confrontation between liberals and conservatives over the loyalty of the Army and which often suggests that the Army was poised on the brink of a *coup d'état*. The Republic, according to such accounts, was saved only when its supporters rallied the masses to save it from a conservative-military takeover. Such interpretations are badly overdrawn and will not hold up under close examination. As a number of recent studies have quite properly pointed out, the French Army—and, in particular, the officer corps—subscribed to the belief that the Army was the "*grande muette*," that is, the obedient and silent servant of the state, pledged to obey the commands of the government that it served. Indeed, this concept of objective loyalty, itself the product of the growing professionalization of the military service, had started to establish firm roots even before the time of the Third Republic. Notwithstanding the rhetoric of the republicans, many of the French bourgeoisie realized this, especially when they found that the

Army could be counted upon when it was employed by the government in labor disputes. Moreover, the loyalty of the Army was never really involved in the Dreyfus case itself. The concept of objective loyalty was already so well established that when, as in the Boulanger affair some years earlier, a few officers did whisper the siren song of military intervention in politics, they found almost no takers within the military hierarchy. In point of fact it wasn't very long after Captain Dreyfus was freed before the republicans themselves wanted to bury the last bones of the affair and make their own peace with the Army. For, as France moved into the twentieth century, the middle-of-the-road republicans began to discover that they feared the generals less than the socialists. And insofar as the socialists were making great political capital over the injustices done to Captain Dreyfus, republicans were moved to restore good relations with the Army. Thus, in many ways the famous Dreyfus case illustrates not the infidelity of the Army but, rather, its loyalty. But, for my purposes, it illustrates one of the central points of this essay: namely, that the history neither of the French Army nor of French national security policy can ever be separated from the sometimes frantic political processes of the Third Republic.

French national security policy, as we have previously seen, rested upon the creation of a mass, conscript army. It was also buttressed by a system of international alliances that began to grow up in Europe in the decade after the Franco-Prussian War. Volumes have been written about the alliance system that began with the Austro-German Alliance of 1879 and that culminated, in the first decade of the twentieth century, with the Triple Entente of France, England, and Russia versus the Triple Alliance of Germany, Austria-Hungary, and Italy. The system is too complicated for any detailed analysis within the framework of this paper. A really appropriate discussion would have to include, for example, such matters as the unintended side effects of particular alliances, the extent to which (as in the case of the Franco-Russian entente) economic rather than military considerations may have been a basic consideration, and whether or not an ally might not actually be a source of weakness rather than strength. But such considerations remain beyond this discussion, so let us limit ourselves to a few observations about the relationship of the European alliance system to French national security.

Much of the literature on the alliance system is focused on the debate over whether or not it was the cause of World War I. That is, was 1914 the inevitable result of an interlocking, competitive alliance system that produced war instead of security—indeed, bred the very insecurities that resulted in conflict? Most historians would agree with this proposition. But in recent years a number of writers have argued, with considerable persuasiveness, that Europe's enjoying under the alliance system almost four decades of peace was a not inconsiderable achievement considering the normal frequency of European conflict. Their more substantive point is that it was not the alliances *per se* that caused war in 1914 but rather changes in the balance of power and in the relative power of particular nations that could not be accommodated within the existing alliance structure.

Yet several conclusions seem inescapable. First, these alliances were part and parcel of the rigid military systems of the participating nations. They were, indeed, primarily military alliances and called for the signatories to produce so many troops by so many days after mobilization had occurred. And, of course, the national armies were rigidly bound to intricate mobilization schemes—literally tied to the railroad timetables. The machinery for putting a mass army in the field was so tightly constructed that, in a very real sense, mobilization meant war. The complexities were such that, once mobilization was ordered, it was virtually impossible for it to be undone, halted, or diverted. Added to these factors was the fact that these alliances were interlocking; once a particular nation ordered mobilization, her allies must also begin to mobilize, and this, in turn, would trigger the response of her enemies. Almost overnight the inexorable machinery would be in motion.

The lack of flexibility in this system of national security can be illustrated by many examples. For example, consider the famous German war plan of 1914, the Schlieffen Plan. That plan had been carefully drawn up as a device to keep Germany from being caught in the middle of a two-front war. It called for the Germans to mass their army in the West and then, by invading Belgium and concentrating the great weight of the German forces on the right wing, to crush the French—above all, to overrun the French Army before the slower mobilization of the Russian Army could be completed. Everything depended upon timing, precise timing, timing that provided no options. France had to be defeated in a matter of weeks before Russia had a chance to put her countless millions into battle on Germany's eastern frontier. The thought that always haunted German military leaders and diplomats was that if there were any delays—military delays, diplomatic delays—then the German Army might be caught in the middle of France with unchecked Russian hordes advancing from the East. The very existence of the Schlieffen Plan was an inhibiting factor in German diplomacy throughout the entire decade before the outbreak of the war. For German diplomats well knew that their nation's one and only war plan called for an immediate invasion of Belgium, the concentration of by far the greater portion of Germany's armed might in the West, and the defeat of France before Russia could get started. Then, when the crisis of the summer of 1914 occurred, the Germans felt that they could not afford the luxury of waiting to see how events developed, of playing for time, or, above all, of letting the Russians mobilize first. And, indeed, when it did turn out that the Tsar was the first to call up his troops, from the German point of view there was no alternative but to act. The logic of the Schlieffen Plan dictated immediate mobilization before Russian action could destroy the German military timetable.

But the Russians were themselves caught in a comparable dilemma of rigid planning. All Russian mobilization plans called for the Tsar's armies to be mobilized against both Germany and Austria-Hungary. Worse still, there were no alternative plans, no plans, for example, to mobilize against one potential enemy and not the other. But in 1914 the Tsar at first wanted to mobilize only against Austria. To have some room for diplomatic maneuvering with Germany, he wanted to leave things militarily loose on the German front. But this was simply impossible. It was not merely that the Russian staff couldn't imagine a war against just one of the Central Powers; rather, it was just not possible, in the absence of alternative plans, to carry out any mobilization except against both Germany and Austria. And so it was done. But, as we have just seen, Russian mobilization was an intolerable threat to Germany, an action that would destroy Germany's one and only plan for waging war. So Germany mobilized, and the wheels were set in motion. In the summer of 1914 the guns of August began to roar because all Europe had adopted national security systems that had no flexibility in them and lacked options or alternatives once the initial step was taken.

The likelihood of conflict was also increased by the general acceptance of war. Before 1914 the European nations regarded it as an acceptable way to solve political problems. Moreover, throughout Europe—and the French were no exception to this—it was generally believed that any war fought by mass conscript armies would be both decisive and short. No war, the experts believed, could possibly last more than six months at the most. Moreover, all of Europe's military leaders placed great faith in the offensive capabilities of their armies. The French, indeed, had carried this faith to the extreme. Their leading theorist, Colonel Loiseau de Grandmaison, developed the concept of the *offensive à outrance* and clearly believed that the offensive spirit of French troops would always triumph over mere firepower. Men like Grandmaison, for example, opposed the development of heavy artillery on the grounds that it hampered the mobility of troops on the offense and nurtured the false doctrine that firepower was more important than spirit and *élan vital*.

But Europeans also expected any general war to be short

because they simply did not believe that any nation's political or economic system could withstand the strain of protracted conflict. They anticipated a collapse of the home front if the war in the field was not speedily concluded; it just did not seem possible that a nation could continue to function if millions of its young men were mobilized and on active duty. And this, by a curious kind of inverted logic, further tended to reinforce the belief that, therefore, any war had to be a short war. Some military writers, to be sure, had given thought to the question of organizing the nation's economic potential for war. But the emphasis had been upon the preparations to be made before war occurred and, at the most, to the efforts necessary to sustain the initial mobilization and preparations for the first great battles. The nations had done this within the short-war frame of reference and had paid no attention to the problem of economic mobilization for a war that might continue over months and years. Thus the French had impressive plans for the military use of the French rail network. But they were designed to speed the mobilization and move material to the areas of deployment; the work was to be completed in ninety days; and when the three months were over, the railworkers were to leave their jobs and report to their regiments at the front. Similarly, there were plans to intensify the production of guns and shells when war broke out. But these were also drawn on a ninety-day basis and, indeed, really amounted to little more than plans to increase the size of the stockpiles that would be available for the immediate climactic battles that were to decide the war. Again, when the ninety days had passed, the workers' exemptions expired, and they were to hasten off to their regiments.

No European nation's national security system in 1914 was prepared for the kind of war that actually developed after the Marne. None anticipated that the weapons of twentieth-century war would make the defense all-powerful; none imagined that the state would be able not merely to sustain itself but even to increase its powers over its citizens and its resources; and none foresaw that the key to eventual victory would be found in the economic and industrial mobilization of all the nation's resources in order to produce the vast amounts of matériel required by the modern military machine.

After 1918 French national security policy would focus precisely upon those things that had been neglected or never anticipated four years earlier. But, above all, what the French learned from the First World War was that they could never hope to stand up alone against a Germany that had recovered her military power. This awareness was the starting point for all national security policy. If Germany managed to escape from the restrictive clauses of the Versailles Treaty, then, all Frenchmen realized, there was no way that France, with her industrial, economic, and demographic inferiority, could match German power. It was a realization with many implications for both French foreign and military policy. At the time when the Versailles Treaty was being written, the realization led French political leaders to insist upon a harsh peace that would "keep Germany down." In the twenties it was the source for the line of conservative, nationalist policy that led to the occupation of the Ruhr and the hard-line insistence that Germany must never be permitted to escape from any of her treaty obligations.

It also led in the diplomatic area to the belief that French security could only be truly guaranteed by a treaty of alliance among France, Great Britain, and the United States. Failing to win American approval, the French then turned to the idea of an Anglo-French security pact, a goal which they pursued throughout much of the 1920s. But Britain and France operated on different wave lengths. Whereas Raymond Poincaré and the French nationalists wanted to treat Germany harshly, the British believed that the political and economic health of Europe depended upon German economic recovery and the political viability of the Weimar Republic. Since their differences over Germany were acute, the French and British could reach few accommodations, and no Anglo-French security pact ever emerged. In their never-ending search for national security, the French, as the next alternative, developed a series of alliances with the new and small states of Eastern Europe—with the Poles, the Czechs, the Rumanians. In the 1920s, when Germany was weak, these alliances looked formidable and helped to give the illusion that France had established a viable system. But in the 1930s, when Russia recovered and Hitler rose to power, the French system collapsed. Indeed, it fell to ruins when the first, faint strains were placed upon it.

But in the 1920s France seemed to the outside world to be a militarist country. Her anti-German policy, her occupation of the Ruhr, her constant attempts to build an Eastern alliance system—these convinced thousands upon thousands of Americans and Englishmen that France was militaristic if not aggressive and unwilling to make the compromises necessary to bring Germany back into the European world. Which is to say that French policy helped to create the opinion, widely held in Britain and America, that Adolf Hitler was only trying to right some of the wrongs imposed upon Germany. Moreover, these French policies heightened the split between the French and British approaches to Germany. They not only destroyed the possibility of creating an Anglo-French security pact but also made it virtually impossible for the two countries to develop a common and consistent German policy. Herein was one of the root causes of World War II: the inability of the two victors of 1918 to work together in the decades from 1919 to 1939. When the two countries that had the greatest stake in the maintenance of peace, in the preservation of the Versailles structure, and in the solution of the German problem could not operate in harmony, the result was both the collapse of the Versailles system and the outbreak of another war.

The second lesson that the French drew from their 1914–1918 experience was that war had now become total war and that plans to fight such a conflict must be drawn years in advance. National security clearly demanded that the nation never be caught as unprepared for total war as she had been in 1914. In the twenties the French—and, again, they were not alone in this belief—devoted great energy to the preparation of plans for complete national mobilization in time of war. Assuming that any future conflict would repeat the pattern of 1914–1918, they drafted detailed legislation to cover every possible aspect of the mobilization of French economic, financial, and human resources for another grinding war of attrition.

In the twenties the planners of General Staff enjoyed much nonmilitary support in this endeavor. Eventually they produced a bill with the somewhat grandiose title, "Bill for the Organization of the Nation in Time of War." Initially, in the mid-twenties, it had strong endorsement. All the French political parties (except the Communists, who were still a tiny minority group in the Chamber) supported it. Socialists and conservatives joined in patriotic praise of the measure. But then things started to go wrong. The business community, aided by certain conservative politicians, began to have second thoughts about "socialistic" attempts to limit wartime profits, impose price controls, and limit the investment of capital. They began to suggest that the proposal ought to be referred back to committee without action. Before long the conservatives were joined by the socialists. The socialists found themselves under fire from the French labor movement, above all, from the Confédération Générale du Travail, the CGT, which wanted no restrictions on wages but insisted that capital and industry should be even more strictly controlled so that "the profit could be taken out of war." The net result was that the momentum for the proposed legislation faded away. By the late twenties the bill had been quietly buried in a legislative committee where it lay dormant for an entire decade. It wasn't until 1938, not until the Munich crisis hit France and Europe, that the bill for the organization of the nation in time of war was suddenly resurrected from committee, brought to the floor of the Chambers, and voted into law after the most cursory of debates.

But this, in turn, merely created some new problems for French security. For in 1938 the French legislators simply reacted to a crisis. They didn't ask the right questions, probe very deeply, or really trouble to find out if a law prepared and drafted over a decade earlier still met the problems that France now confronted. They simply dragged it out and plugged it in. Some, though certainly not all, of France's problems with wartime production were related to this. Likewise, in 1936 and 1937 the French had nationalized their armaments industry. Supporters of nationalization have argued that this was a measure designed to maximize armament production for a war which even then was beginning to be visible on the horizon. But in point of fact, most of the pressure for nationalization of the armaments industry came from the French Left, which wanted to make sure that no private arms manufacturer would profit from war. The French legislation was part and parcel of the broad movement against the so-called "merchants of death" that was prevalent in Britain and America as well. And this was the underlying motive rather than any farseeing plan to rationalize armament production in the interest of national security.

Needless to say, many of these observations remain controversial. The issue of French war production—how many tanks and planes were produced, the actual French deficiencies in air and mechanized power—has caused great controversies. Both the Vichy Government and the Fourth Republic staged highly publicized trials and investigations to unearth the "guilty" and find those presumably responsible for shortages of French armaments. It is no trick at all to find a whole series of French political leaders who served the Popular Front governments after 1936 and who maintained steadfastly that their industrial mobilization plans worked quite well. The defeat of 1940, they claim, was not the fault of the politicians for providing insufficient matériel, but the failure of the military leaders to understand how to fight a mechanized war. Their view is challenged by an equally assured group of conservatives, with allies from the ranks of the military, who contend that France's lack of preparation started with the social reforms of the Popular Front, with the nationalizations and, especially, with the forty-hour week. These alleged reforms, so runs the conservative complaint, weakened war production to such an extent that in 1939

France went unprepared to her military doom. There is not space to attempt any definitive resolution of these differences of opinion—even if we make the questionable assumption that such a resolution is really possible. However, the balance of the evidence favors the men of the Popular Front and not their conservative enemies. In 1940 France had as many tanks as the Germans, and her deficiencies in aircraft were not overwhelming. What was wrong was the way they were employed and the kinds and types of tanks and aircraft that the military had asked for. The real failure was in French military doctrine.

A third factor that affected French national security policy in the interwar years was the widespread popular desire, pronounced in the 1920s, to reduce the length of military service. By 1928 this demand led to the reduction of a conscript's tour of duty from two years to twelve months. Such a reduction in the length of service was, to be sure, also made possible by the fact that under the Versailles Treaty the German Army had virtually ceased to exist. It was, you will recall, limited to 100,000 professionals, and even to the most alarmist of the French, such a force posed no clear and present danger. But there were other important reasons for the change in length of service.

Military writers, as we have seen, assumed that a future war would take the general pattern of the trench warfare of 1914–1918—that is, they foresaw a war of attrition, of little movement, and in which the defense would clearly overmatch the offense. Under such military assumptions, then, there would clearly be time to organize the defensive strength of the French nation and mobilize national resources. It also followed from these assumptions that a "defensive army"—an army built out of a relatively small cadre of professionals but composed primarily of men serving a year's tour of duty—could hold the line while the reserves were being called up and the industrial wheels made to spin faster. Thus belief in the 1914–1918 experience as the model for future war led logically to acceptance of a greatly reduced length of military service. And, in point of fact, under the one-year law the army in being—the so-called "active army"—became little more than a training force whose principal function was the military instruction of each annual class of raw recruits.

Closely associated with the 1928 conscription legislation was the construction of the Maginot Line, that famous (or is the proper word infamous?) system of concrete fortifications along much of France's exposed eastern frontier. In fact, the one-year service would not have been accepted without the concurrent decision to build the Line. For while the Staff did accept the idea of a defensive, semi-static war in which an army of one-year conscripts and reservists could adequately protect the nation, they also believed that national security required a system of fortifications that would reinforce the defense and, above all, provide the time necessary for effective mobilization of the reserves.

The Maginot Line has received blistering criticism over the years. But at the time it was built and in terms of the frame of reference of the late 1920s and early thirties, a logical case can be made for it. The Maginot Line was not intended to be a Chinese wall that would hermetically seal France from the rest of Europe. It was simply intended to reinforce the defense—to blunt any immediate German attack and, by so doing, to provide the nation with the time necessary to mobilize its total defense machinery. The trouble arose in later years—in the 1930s—when political and military leaders began to reverse the argument. They began to argue that the existence of the Maginot Line made it unnecessary to carry out reforms or make changes in France's military posture. One began to hear the argument that ran, in effect, "Since we have the Maginot Line to protect us, we therefore don't need to lengthen the time of military service . . . increase the number of tanks . . . create armored divisions, etc., etc." Which is to say that, with the passage of time, the existence of the Maginot Line became a justification for standing pat, for rejecting innovation. Moreover, the later exaltation of the Maginot Line furthered the tendency of Frenchmen to view war in static and defensive terms.

In any event, by the early 1930s the French had built a

system of national security founded on principles that were almost completely the reverse of those that had prevailed in 1914. But this system, too, was inflexible and proved unable to respond to crises short of total war. In 1936 the French were faced with the first real German challenge to the Versailles system. when Hitler moved into the Rhineland. It is now well known that Hitler dispatched only a few regiments and that, from a strictly military point of view, the move was both gamble and bluff. What happened in Paris was, roughly, as follows: the leaders of the French Army informed the Cabinet that French military forces were not organized to carry out a rapid movement of their own into the Rhineland and could not, in short, mount a quick surgical operation. Indeed, the high command suggested that any sort of French military response would require something close to a national mobilization. Worse still, they sketched out a scenario in which there was a real prospect that any French operations in the Rhineland would bog down in a trench warfare stalemate on the 1914–1918 model. In other words, when French military leaders were consulted about the Rhineland, they could only recommend a fairly complete mobilization, could not promise any quick, surgical response, and had to confess that any military venture might turn into a stalemate. The government in power in March of 1936 was strictly a caretaker regime, a cabinet minding the political store until national elections in June. It is no wonder that, receiving such advice, they decided not to act.

Other comparable examples could readily be found, but the Rhineland incident stands out for the way that it revealed the limitations of the defensively oriented national security system that the French had created. Moreover, the revelation of such limitations served only to further appeasement. Within a short time leaders of the French Right were arguing that since the nation did not possess any army capable of pursuing an anti-German policy, then the nation should pursue a foreign policy that was in accord with her military situation: that is, a policy of appeasing Hitler and acceding to further German moves.

It is by no means fair, however, to blame this situation solely on the military hierarchy and its defensive mentality. French political leaders in the 1930s accepted the prevailing military system and subscribed to the same theories of warfare. It would be quite possible to construct a series of parallel quotations that would match statements by military leaders on the virtues of the defense with similar comments by members of the political leadership. Even as late as Munich, Edouard Daladier, the Prime Minister, told his colleagues that the experiences of the Spanish Civil War clearly demonstrated that a few soldiers armed with machine guns could withstand heavy frontal attacks even when these attacks were supported by armor. A few individuals, to be sure, challenged both the military system and the theories on which it was based. The most prominent was Charles de Gaulle. In the mid-thirties de Gaulle developed his own military prescription in a short book, *Vers l'armée de métier*. De Gaulle demanded the creation of an entirely new military structure; his book was a call for France to break out of the defensive mold in which her national security system had been formed. He advocated the creation of a professional army that would be equipped with mechanized divisions and air power, a striking force with sufficient power and mobility to restore the offensive to war.

To what extent was de Gaulle a prophet without honor in his own country? Two conclusions clearly emerge. First, de Gaulle was an enthusiast, a man with a cause, an advocate who never conceded that his offensive, striking forces could be checked. It was thus relatively easy for the leaders of the Army to write books and articles of their own in which they argued that, unlike de Gaulle, they took account of the basic need for a balanced military system and that it was highly dangerous to overemphasize any one component of the national defense, such as air or mechanized forces. De Gaulle, the hierarchy maintained, asked France to place all her military bets on a single, one-sided solution and failed to comprehend the need for balanced forces. But, above all, the principal reason why de Gaulle was unsuccessful was that he placed so much emphasis upon replacing the conscript army of citizen-soldiers with a strictly professional force. De Gaulle, always something of a technocrat, maintained that modern weapons were so complex and sophisticated that draftees, in service for only a short period of time, could not learn to use them effectively. Only professionals, who devoted their lives to the Army, could develop the necessary expertise. But it was still impossible to receive political support for the idea of a professional army in republican France. French political leaders regarded the proposal as unthinkable, and believed that to support it would be tantamount to committing political suicide. One prominent Frenchman, Paul Reynaud, did put de Gaulle's ideas before the Chamber of Deputies in 1937. The debate on Reynaud's bill simply proves the point. There was virtually no discussion of the military merits of the de Gaulle proposals or his views on the role of air and armor. What happened was that the debate focused on the political implications of a professional army, with speaker after speaker rejecting the concept as anti-democratic and foreign to the republican tradition. The basic reason for de Gaulle's failure was that he hitched his air-armor chariot to the idea of a professional army, and the latter was simply not acceptable.

Finally, the national security policies that France pursued in the 1930s cannot be considered apart from the complicated and often bitter political history of that decade. The ultimate failure of French security policies in 1940 was to a considerable extent determined by internal political developments.

First, there was a truly striking reversal in the traditional role of the leading political parties. Up until the 1930s the great defenders of the Army and the strongest advocates of a firm anti-German security policy had been the conservatives and the nationalists. But between 1934 and 1936 these groups began to change their outlook. Conservatives, and, indeed, even center and middle-of-the-road political groups became increasingly afraid of domestic social disorder and the prospect of sweeping social and economic change. They were positively frightened by the great riots in Paris in February of 1934, by the spread of industrial strife, and, above all, by the rise of the Popular Front—an alliance of the French Left that included the Communists. When the Popular Front won a sweeping electoral victory in the spring of 1936, fear turned to apprehension and dread. Conservatives and nationalists now began to look upon Adolf Hitler not as the enemy of France but, rather, as the defender of middle-class, conservative values against the progress of Bolshevism. The extreme form for this conservative reaction was expressed in the slogan, "Better Hitler than Léon Blum."

The foreign policy views of French conservatives naturally shifted. Abandoning their traditional anti-German perspective, they now began to advocate what eventually became known as appeasement. They moved surely and steadily in the direction of finding some form of accommodation with Hitler. They used military arguments to support their case for appeasement. Conservatives began to point to the defensive orientation of the French Army and the French inability—once the Rhineland was reoccupied and the Siegfried Line under construction—to support her eastern allies or sustain an anti-German policy. Realism, they maintained, dictated a foreign policy of accommodation that was in accordance with French military posture. But the underlying motivation was their domestic concern, their fear that the Popular Front was putting France on the road to socialism.

This reversal of roles helps to explain why the French moved so uncertainly on the international stage after 1936. Appeasement was one consequence of the larger problem of a society polarized by political disagreements. Both foreign policy and national security issues were gravely weakened by these great unresolved domestic concerns.

Another important factor in the ultimate breakdown of French national security policy was the increasing disaffection of the officer corps in the decade of the 1930s. This, to be sure, is a complex issue and one on which it is dangerous to be dogmatic. But, in the pre-1914 era the French Army was genuinely loyal to the Republic and prided itself on serving as the silent, obedient defender of that state regardless of the nature of the regime in power at the moment. Yet this attitude, too, began to change in the decade before World War II. The officer corps was deeply disturbed by drastic cuts in the military budget, by the infamous Stavisky scandal (which suggested that numerous public officials were corrupt) and by the 1934 riots in Paris. The rise of the Popular Front led to increased disaffection and a politicization of significant elements of the military hierarchy. It is, of course, true that the Third Republic was still strong enough to appoint a thorough loyalist like Maurice Gamelin to the command of the French armies and to sustain him in that position. And Gamelin was indeed the perfect model of an Establishment figure, a defender of the regime through all adversities. But Gamelin was no longer truly representative of the political spirit of the officer corps, especially its top echelons. Behind him were the Pétains, the Maxime Weygands, and other senior officers who not only distrusted the regime in power in Paris but were also becoming disenchanted with the Third Republic itself. Their alienation emerged fully in June, 1940, when the French armies were being crushed by Hitler's *Blitzkrieg*. Their decision to surrender was directly tied to their political views, above all, to their disaffection with the Republic. When Pétain and Weygand advocated surrender and refused to continue the fight from North Africa, their recommendations were in large part conditioned by their belief that accommodation to Hitler's New Order would ultimately lead to a restoration of lost civic virtues and provide guarantees against the communists and leftists who, in the eyes of these officers, had undermined the Republic.

The disaffection of important elements of the officer corps became an important strand in French military history after 1940. Indeed, after 1940 one cannot write the history of the French Army without focusing upon a new and now often central issue: the extent of the loyalty of the Army to the State. Indeed, the very history of the Fourth and Fifth Republics turns upon such matters as the anger of officers who felt betrayed by the defeat in Indochina or the machinations of the colonels in Algiers who were prepared to destroy the state to prevent Algerian independence.

All of which leads back to the central premise of this paper: that French national security policy has to be considered in terms of broad political, economic, and social phenomena. I myself have always been most impressed by Marc Bloch's book *Strange Defeat*, a study of what went wrong in 1940 and, in my opinion, a classic example of this broad approach. Bloch was a professional historian—indeed, one of the finest medievalists of his time—but he served as a staff officer in both world wars. *Strange Defeat* was written in anguish within a few months after the French surrender; it is a first-hand, intensely personal, often subjective account by an eye witness. One of the most devastating condemnations of the French General Staff ever to be written, it charges that the military leaders of 1939–1940 failed because they based all their planning on the tempo of 1914, on the speed and mobility of the soldier on foot. They were, Bloch attested, intellectually and psychologically unprepared for the reality of the *Blitzkrieg*. Bloch was one of the very first writers to capture the essence of what has since become a commonplace: that the weapons developed after 1918 revolutionized war by compressing both time and space. And he was also among the very first to portray the psychological shock—shock that paralyzed action—of an officer corps that suddenly found itself plunged into an unexpected military situation and lacked the intellectual preparation to cope with it.

Most readers stop at this point and read no further. But Bloch was too careful a historian to believe that he had provided a complete explanation for the French military disaster by pointing to the failure of the General Staff and the military hierarchy. The latter portion of *Strange Defeat* is a perceptive critique of the society in which he lived, above all, of the mores and beliefs of the dominant political and economic groups. He spelled out the unwillingness of the middle classes to carry out the internal economic and social changes that were necessary to make

France a viable nation. His ultimate verdict was that, in the last analysis, France in 1940 committed suicide—for the defeat was the responsibility of all Frenchmen and not just of a few staff officers who had outdated ideas about the way to fight a modern war.

It seems to me that the basic problem of French national security in both wars was that they were too wedded to particular organizational forms and too committed to certain patterns of thinking about the nature of war. On both occasions their system lacked flexibility. The French were simply too set in their beliefs, too certain that they had learned the proper lessons from previous military experiences. But perhaps we Americans should not be excessively critical.

Americans have often tended to think that the French have a sorry military record in recent years and that, somehow or other, we can do better. Yet in 1940 the French were the first to face the *Blitzkrieg,* and it is by no means certain that other nations, placed in the identical situation, would have responded any more imaginatively or successfully. Indeed, there is a certain danger in assuming that French military ineptitude provides opportunities to demonstrate American prowess. Several years ago I helped to conduct an oral history project at Princeton in which we tape-recorded the recollections of men who had known and worked with John Foster Dulles. Since Vietnam was already a central issue of American policy, we naturally asked a lot of questions about how the United States had gotten involved in Indochina during the Eisenhower-Dulles years. I myself encountered several American officers who started from the assumption that the French had botched their operations in Vietnam and would not have come to the disaster of Dien Bien Phu if they had listened to American military advice. It was clear to me that these American officers felt that the North Vietnamese and Vietcong had succeeded not so much through their own efforts but as a consequence of French blunders. They also believed that Americans, if and when they became involved, would never make such mistakes. In a very real sense, then, the history of the American involvement in Vietnam is, if nothing else, an object lesson in the danger of making too many generalizations about French military failures and being too dogmatic about the "lessons" of the French military experience.

Bibliography

For further discussion of many of the issues raised in this paper, see my book *The French Theory of the Nation in Arms, 1866–1939* (New York: Russell, 1955).

The best studies in English on the French Army are Paul-Marie de la Gorce, *The French Army: A Military-Political History* (New York: Braziller, 1963) and David Ralston, *The Army of the Republic: The Place of the Military in the Political Evolution of France, 1871–1914* (Cambridge: M.I.T. Press, 1967). In French the most helpful analysis remains, Raoul Girardet, *La société militaire dans la France contemporaine* (Paris: Plon, 1953).

On the alliance system and mobilization in 1914 see Gordon A. Craig, *The Politics of the Prussian Army* (New York: Oxford University Press, 1955) and Luigi Albertini, *The Origins of the War of 1914*, 3 vols., (New York: Oxford University Press, 1953–1957), especially volume II.

The best study of the loyalty of the officer corps to the regime and the Republic is Philip Bankwitz, *Maxime Weygand and Civil-Military Relations in France* (Cambridge: Harvard University Press, 1967).

The political reversals of the middle and late 1930s are best analyzed in Charles Micaud, *The French Right and Nazi Germany, 1933–1939* (Durham: Duke University Press, 1943).

On French foreign policy in the 1920s, see my article "The French Foreign Office: The Era of Philippe Berthelot," in Gordon A. Craig and Felix Gilbert, eds., *The Diplomats* (Princeton: Princeton University Press, 1953). On the 1940 surrender, see my chapter "The Third Republic and the Generals: The Gravediggers Revisited," in Harry L. Coles, ed., *Total War and Cold War* (Columbus: Ohio State University Press, 1961).

Charles de Gaulle's plan for a professional, mechanized army has been translated as *The Army of the Future* (London: Hutchinson, 1940), while there is an English translation of Marc Bloch's book under the title, *Strange Defeat: A Statement of Evidence Written in 1940* (New York: Norton, 1968).

COMMENTS

Samuel P. Huntington argued in his influential study The Soldier and the State: The Theory and Politics of Civil-Military Relations *(Cambridge: Harvard University Press, 1957), that the problems of civil-military relations in Japan were not those characteristic of the Western great powers after the rise of Prussian military professionalism, but rather were unique. In imperial Japan, Huntington contended, a true military profession in the Western sense never developed. The Japanese officer was not a professional governed by the distinguishing characteristics of professionalism in the West, whether of military professionalism or of any other kind: learned expertise, responsibility to the whole society, and corporateness. Japan adopted only the external shell of Western military professionalism. The imperial Japanese officer was not a professional soldier but a warrior— "a fighter engaging in violence himself rather than a manager directing the employment of violence by others. This was a feudal, not a professional, ideal." (p. 126). Because Japanese officers were not military professionals in the Western sense, however, Huntington argues that the Western tension between the soldier and the civilian did not develop in imperial Japan. Rather, the warrior code of ethics of the Japanese military was also the national ideology of Japan—the Bushido code. In Huntington's view, Japan followed a very different road to military defeat than did Germany in 1918 and 1945 or France in 1940.*

Alvin D. Coox of San Diego State University has written extensively on Japanese military history, including a number of articles and Kogun: The Japanese Army in the Pacific War *(in collaboration with Hayashi Saburo; Quantico: Marine Corps Association, 1959). In his lecture of 1973, Coox did not directly challenge Huntington's interpretation, but his implication is that relations between the soldier and the state in Japan were less different from Western patterns than Huntington would*

=== COMMENTS ===

have them. While Coox acknowledges that Imperial General Headquarters "bore only a surface resemblance to the Western type of JCS organization," he is speaking in this instance of unification of the various armed services, not of the essentials of military professionalism and military values or of civil-military relations. In these essentials, Coox suggests that Japan offers another case study of problems common to all the modern great powers.

The Japanese Army Experience

ALVIN D. COOX

Last Days of the Samurai Caste

ONE hundred and twenty years ago, Commodore Matthew Perry first visited Japan. The country was still under the temporal control of the *Shogun,* feudalistic generalissimos from the House of Tokugawa who administered the land from Yedo, modern Tokyo, ostensibly in the name of the Emperor. The latter resided in gilded impotence hundreds of miles away at Kyoto. There was no national army, only regional forces belonging to the clan overlords, the *daimyo,* whose military vassals were known as *samurai* or *bushi*. These were the warrior-administrators who had dominated Japanese society for seven centuries before Perry's Black Ships arrived in the bay of Uraga.

It was still the time of the brass muzzle-loading cannon and the two-handed sword and dagger. Perry was singularly unimpressed by his first sight of the Japanese military. The *samurai* swords seemed to be "more suited for show than service" and "awkwardly constructed for use." As for the 5,000 Japanese troops ashore, their "loose order":

> . . . did not betoken any very great degree of discipline. The soldiers were tolerably well armed and equipped. Their uniform was very much like the ordinary Japanese dress. Their arms were swords, spears, and matchlocks. Those in front were all infantry, archers, and lancers; but large bodies of cavalry were seen behind . . . . These troopers, with their rich caparisons, presented at least a showy cavalcade.

NATIONAL SECURITY AND MILITARY COMMAND

In all, Perry judged that his 300 sailors, marines, musicians, and officers, although "no very formidable array [were] composed of very vigorous, able-bodied men, who contrasted strongly with the smaller and more effeminate-looking Japanese."

The more astute Japanese leaders were well aware of the nation's weaknesses. As a Japanese police official reported on the American landing party, "[We] knew we could not control these people. We had to hold our anger . . . ." The more impressionable among the populace were so cowed that they estimated the personnel strength of the ten-ship American expedition at 100,000 officers and men. Perry obtained his treaty, and Townsend Harris, the unwanted consul, arrived not long afterward. Within the decade, when Japanese hotheads at the "Impregnable" forts of Shimonoseki and Kagoshima dared to engage foreign warships, they felt the fierce edge of Western military technology in the breech-loading British naval cannon, and their xenophobia was converted into professional emulation.

This was the period of *Bakumatsu* or the end of the feudal era. When the last *Shogun* gladly surrendered his secular powers to the teenage Emperor, there ensued the exciting period usually known as the Meiji Restoration. The motto of the Japanese modernizers (who indulged in slogans quite as often as the propagandists of the People's Republics today) was Military Strength and National Prosperity. As Yamagata Aritomo, a founding father of the Japanese Army, put it as early as 1871: "In view of the prime consideration on making provision against Russian penetration to the south, armaments should be expanded, and . . . should be given priority over all other policies." Japanese historians see, in this early concept of the Army, the "leading function of uniting the whole of Japanese society under the principles of militarism." Yamagata was explicit as to his meaning, however:

> An "open and progressive policy" does not mean simply open ports and commercial profits. It means also that Japan must promote her national prestige, in order to maintain a

dignified and independent position among the powers . . . . In my opinion, the most important and urgent business is the increase of the defensive forces, whereby the very existence of the nation can be safeguarded and her perfect independence maintained . . . . Although this will require great sums of money, and so increase the burden of the people, it is an inescapable responsibility. The impoverishment caused by additional military expenditures may seem a national misfortune for the moment, but in the end it is the only means by which to secure our national welfare . . . . For what should have been during the past hundred years, we are now obliged to do in the next decade.

The first priority was to centralize and solidify the political system, and to liquidate the *samurai,* figuratively, as an elite military, social, and economic caste. This was a dangerous business for, in the first months of the struggle between forces loyal to the Throne and those clinging to the Shogunate system, the Emperor could directly control only a motley collection of some 300 warriors recruited from the lesser clans, in an ocean of perhaps 400,000 dispossessed *samurai*. Under the circumstances, the tasks of the Meiji government would be accomplished more by purchase, encouragement, and financial guarantee than by naked threat.

Early Military Structure

Organizationally, in the first two years of the Meiji era, the government was still administered by state councillors. Then, as the machinery grew more complicated, larger, and more specialized, the administrative organs were reorganized frequently. In the military sphere, an Army and Navy Affairs Section became the Office of Military Defense and then (in June 1868) the Department of Military Affairs. Next year, government departments were established, including a Ministry of Military Affairs, encompassing the Army and the Navy. In 1871 an Army Staff Bureau was set up, on the French pattern, "to take part in secret duties and planning, collect maps and political information, and be in charge of spies, intelligence, and other matters." (There was no Navy counterpart till 1887). The chief of bureau was also vice minister, and thus entirely under the ministry's control. But, significantly, this officer—the bureau chief —had to be of the rank of colonel or above, while the minister had to be a major general or higher. Thus, so early in Japanese army history, was the system of exclusive appointment of military officers introduced into the governmental hierarchy, ostensibly in order to "give priority to military efficiency."

Upon the revamped sociological base achieved by socializing the classes, the Meiji government proceeded to erect a new, uniform, national military establishment. The clan armies of feudal retainers were disbanded, and the castles were taken over. An Imperial Guard was organized from the great clans, the Tokyo and Osaka garrisons were reinforced, and two forces were assigned to the northeast and to Kyushu. Although we are speaking only of ten battalions, this force represents the first major step to form a national standing army. Never before had a fixed number of troops been stationed in garrison towns "for the purpose of preserving peace and maintaining the social order." Reforms in military organization accompanied the transfer of military power into the hands of the new imperial government.

One of the most important of the military reforms was the institution of country-wide conscription in 1873. "First [we must] prepare against civil disturbances," said the Meiji planners; "later, prepare against foreign invasion." The Japanese would have to complement their "ancient military system with the excellence of the Western, in order to meet the national emergencies with a proper army and navy." Among the several considerations was the fact that a conscription system was seen by the hard-pressed government as far less expensive than that of a permanently subsidized professional army made up of the former *samurai*.

The first, French-tinged draft law imposed a military responsibility between the ages of seventeen and forty with active duty of three years from the age of twenty, and four years' obligation in the reserve. Under German influence, the law was revised in 1889, reducing the generous exemption features and providing for three years of active training, four years of obligation in a first reserve, and five years in a second reserve. Those males seventeen to forty who were on neither the active nor the reserve lists must be enrolled in the militia.

From 1873 the country was divided into six military districts, each with a garrison of 40,000 in peacetime and 70,000 in wartime, centering on an infantry brigade per garrison (Tokyo, Sendai, Nagoya, Osaka, Hiroshima, Kumamoto). This would amount to a regular army of 400,000 in wartime, and certainly exceeded the short-range objective of suppressing "any civil disturbance that might arise due to dissatisfaction among the former retainers of the feudal clans." The Army was already thinking of the day when it could not only store its own ordnance and ammunition but produce them at home.

Obligatory national military service remained the key to the strength of the Japanese Army, despite uneasiness on several sides. There could be no doubt that conscription would undermine or at least prove incompatible with the legacy of elitist *Bushido,* with its unique code of moral discipline and emphasis upon individual skills. The draft proved to be a "powerful force for turning the parochial loyalties of the people into a generalized patriotic feeling for the nation as a whole." To the founders of the conscription system, "In due course, the nation will become a great civil and military university . . . . " Thus the Japanese Army was an agent in combating illiteracy, teaching special skills, and inculcating a pseudo-religious national ideology centering on Throne and polity *(kokutai).*

Specialization of Military Functions

In February, 1872, the Military Affairs Ministry was divided into separate Ministries of the Army (or War) and of Navy. Inevitably, within the services, staff and command authority had to be separated from administrative functions. As a first step, in 1874 a Staff Bureau was created, and placed outside the War Ministry. Influenced by German military institutions and administrative organs, the Japanese authorities set up an independent General Staff Office *(Sanbo Honbu)* in late 1878, confirming the existence of two chains of command, administrative and operational. The Chief of General Staff "shall take part in the secret duties of the military council," that is, assist the Throne in important secret duties at the Emperor's operational headquarters. In time of war, the Chief would pass down operational instructions to commanders after receiving the personal sanction of the Emperor. In practice, Imperial Orders (which always required sanction) preceded AGS Directives (explanatory instructions which did not require sanction).

The Army General Staff was now fully responsible for plans and strategy, but there was a far more significant political effect: As major military adviser to the Emperor in these areas, the Chief of AGS was independent of the War Minister and superior to him. When Yamagata was promoted (perhaps one should say, promoted himself) from War Minister to Chief of AGS, he brought a clan outranking as well as a professional outranking over the new War Minister. In other words, in these early days, more account was taken of the balance of power between the *Choshu* and *Satsuma* cliques than of the later norms of performance and ability pure and simple. These measures effectively removed the military command function from the possibility of political checking by the Cabinet, In fact, a reason for separating the AGS from the War Ministry and for setting the Chief of Staff directly under the sovereign, in the chain of command was to prevent "interference" by the civil government in matters affecting national defense and military strategy.

A closely related step was the assignment of independent status and authority to the military body responsible for training and inspection. In 1879 a Board of Supervision was established; from 1888 a full general or lieutenant general was to report directly to the Throne. Renamed the Board of Military Education in 1898, this training and inspection agency was temporarily allocated to the War Ministry, but after 1900 it returned permanently to direct imperial control.

Thus, at the highest level of the Japanese Army, there was a War Ministry charged with military administration; a General Staff, directly responsible to the Throne for military advice; and an Inspectorate General of Military Education, as it was eventually called. The generals in charge of these important organs constituted a very influential triumvirate known as the Big Three of the Army.

When the first Japanese Cabinet was formed in 1885, the War Minister became the Cabinet Minister representing the Army. Meanwhile, with the gradual development of a navy, the Navy Bureau was set up within the Army General Staff in 1886. In the reorganization of 1888, a Naval General Staff was created, but since naval affairs were still relatively simple, the NGS was moved to the Navy Ministry in 1889, and still came under the Chief of Staff for all imperial armed forces, who was the AGS chief. On the eve of the Sino-Japanese War, in May 1893, the Naval General Staff was finally placed under direct control of the Emperor and made coequal with the Army General Staff.

The Army abolished its old garrison structure in 1888, in favor of six divisional districts and a division of Imperial Guards (from 1891), each division to be commanded by a lieutenant general. According to the official Army explanation of these changes, it was because overseas operations were now foreseen, especially in view of deteriorating relations with China since the recent clashes at Seoul in 1882.

The "Emperor's Army"

The rapport between Throne and services contributed greatly to the strengthening of the imperial system of government. A number of devices were contrived whereby the authorities and the high command could bind the military to the patriarchal monarch, as "His army." Thus Article I of the Army Service Regulations of 1879 stated that the Imperial Army was directly responsible to H. M. the Emperor. And the very famous Imperial Precepts to the Soldiers and Sailors (1882) stated that:

> The Supreme command of Our forces is in Our hands, and although We may entrust subordinate commands to Our subjects, yet the ultimate authority We Ourself shall hold and never delegate . . . . Soldiers and Sailors, We are your supreme Commander-in-Chief. Our relations with you will be most intimate when We rely upon you as Our limbs and you look up to Us as your head . . . . Inferiors should regard the orders of their superiors as issuing directly from Us.

According to the Constitution of 1889, the "sacred and inviolable" sovereign determined the organization and peacetime standing of the armed forces, and possessed sole power to declare war, make peace, and consummate treaties. The juridical aspects of these powers are summed up in the phrases, *Independence of the Prerogative of Supreme Command* and the *Prerogative of Military Administration*. In practice, however, asserts Professor Fukushima Shingo:

> . . . the members of the military forces did not recognize that the Emperor was bound by the Constitution. Consequently they were incapable of understanding theories such as those in which a distinction is made between the exercise of the two capacities of the Emperor, and they believe that, provided the Imperial Prerogative of Supreme Command had been personally exercised, cabinet decisions could be overridden without impediment. If such an interpretation is adopted, it means that the will of the armed forces as represented by those who assisted the Emperor in his exercise of the Prerogative of Supreme Command had, for practical purposes, attained the position of the supreme directing will in the state.

The soundness of this criticism is suggested by the fact that Cabinet documents were countersigned by the Premier and individual ministers, indicating their reponsibility. But military orders emanating from the high command bore no countersignature and therefore represented absolute orders issued by the monarch in his capacity as Commander in Chief. Since 1886, too, membership in the Imperial Family was made prerequisite to service as Chief of General Staff. The most famous feature of the "dualistic" Japanese command structure derives from the requirement that the military and naval members of the Cabinet, the War Minister and the Navy Minister, be active duty general or flag officers. Although this provision was modified, *ca.* 1914, to the effect that the service ministers could be chosen from the reserve or retired lists, this stipulation was never invoked; and, finally, the old active duty requirement was reinstated after the disturbances of 1936. As as a result, the services wielded a life-or-death control over the fate of Cabinets, even in cases when the armed forces should have had no concern with the matters under discussion. As critics put it, a service minister could topple a Cabinet in which he was a minority of one.

The Japanese Army never shook off entirely the characteristics of being a private army of the monarchy, partly by historical accident but largely by design. "Since war and military matters proceeded from the action of the Emperor, they were considered sacred, and to oppose them or to speak of peace was at once considered criticism of the Emperor, although it was permitted to speak of a peace of conquest to be attained through war." By linking all ranks to the divine sovereign, a blind and mystical obedience could be invoked. For, as the councillors stated in a memorial of 1881:

> . . . it may be said that the Son of Heaven is the generalissimo of the forces in the field, and the members of the armed forces are the teeth and claws of the Royal House. Therefore those who are members of the armed forces have the duty to love their country wholeheartedly and to be loyal to their sovereign.

In pursuit of these goals, the Japanese Army provided intensive Spiritual Training or, as one commentator called it, "cultivation of the psyche—the life blood of training, not even to be neglected while sleeping." The ethics textbooks reminded all readers that the Emperor "cherished His subjects as though they were His children." Of the monarch, it was stated that, "The whole nation is Our family and every inch of the country is Our household." Even the lowliest private soldier saw omnipresent evidence of the imperial presence in the royally bestowed regimental colors, in the daily recitations of the Precepts to the Soldiers and Sailors, accompanied by obeisance in the direction of the Imperial Palace wherever one might be stationed, and in the orders of all superiors, from the noncoms up. In January, 1972, we observed amazing evidence of the vitality of this sense of responsibility between soldier and Throne when the last Japanese Army straggler (Sergeant Yokoi, age fifty-six) was found alive on Guam. Most of his old rifle had rotted away but he had faithfully preserved the imperial chrysanthemum crest embossed on the stock, by burnishing it during *twenty-eight years* of skulking and foraging in the island brush.

Military Training: The Matter of Foreign Advisers

Even in the last decades of the Shogunate, it was realized that Japanese levels of military skills, science, and technology were no match for Western armies and navies. Various arrangements were made with the Dutch, English, and French, in particular, to provide not only ordnance but military training. In early 1867 a French mission arrived, under an army captain, to train about 250 infantry, artillery, and cavalry cadets at a new military school near Yokohama. The training grounds were soon moved to the Tokyo area, but the French had backed the wrong side, the Shogunate, in the impending conflict against the Crown. After only a year and eight months, of which only some six months could be devoted to practical training, the French military mission had to be recalled.

With the solidification of the new imperial regime, the Japanese authorities again were obliged to consider the recruitment of foreign military advisers. An internal struggle raged between proponents of the French versus the Prussian systems. Meanwhile a military academy which had been formed in Kyoto was moved to Osaka, under imperial auspices. By 1870, over 400 cadets were in training, on the style of the brief French mission of 1867–1868. Ironically, a month after the French Army was shattered by the Prussians at Sedan, the Japanese government announced the adoption of the French military system. Ernst L. Presseisen has demonstrated that it is merely a legend that the Japanese switched to the German system after France's defeat in 1870; not one German army instructor was hired by the Japanese until 1885.

Lieutenant Colonel Charles-Antoine Marquerie (later replaced by Charles Munier) brought a French army team to Japan in 1872. The military training school had already been moved up to Tokyo from Osaka and, under French guidance, a single Military Academy (the *Shikan Gakko*) was placed in operation from 1875. Top graduates were sent abroad to study; forty-four officers were in Europe in 1880. French military influence, even at the "Japanese St. Cyr," began to wane after the Japanese government's painful field experience in putting down a serious rebellion in 1877, although the French contracts were extended through 1880.

As the French mission's fortunes declined, the prospects for the Germans rose, especially since important Japanese officers who returned from study in Berlin brought back enthusiastic recommendations about the superiority of the German Army. In 1885, Major Klemens Wilhelm Jakob Meckel was hired to teach and to advise the General Staff. The establishment of the Army War College in 1883 led to a downgrading in importance of the French creation, the Military Academy. For graduates of the War College, the road was open to the highest ranks and the most influential posts; those who wore the badge signifying graduation from the War College were an elite group in the army officer corps. The Emperor himself bestowed awards (including an imperial saber) upon the star students.

Meckel's contract was extended to a second and then a

135

third year, through 1888. A second German officer was recruited in 1886. Although there were still four French army instructors in Japan, after 1884 they no longer trained troops. Most of the vestiges of French advice were eliminated when a new conscription act was introduced in 1889. "German ideas [had] conquered the Japanese Army." But the Japanese were tiring of the political and diplomatic disputes engendered by the presence of rival European officers. The French officers were recalled by their government in late 1888; and Meckel's German successor was not extended after his contract ran out in two years in 1890. Thinking that all foreign military advisers were now to be dispensed with by the Japanese Army, the French withdrew their military attaché, a unique post held in Japan for ten years by Captain Alexandre Bougoüin. But the Japanese surprised the French by engaging the services of another German officer for duty at the Army War College. He remained until 1894, the eve of the Sino-Japanese War, by which time the conversion of the Japanese Army to the German model, was completed.

Although the Japanese put their military training to good use in the subsequent wars against China and Russia, they burned their fingers during the Triple Intervention, when Germany teamed up with France and Russia to compel the Japanese to disgorge important acquisitions in South Manchuria. Thereafter, the Japanese Army decided, it would send more of its officers to study abroad, and hire no foreign officers to teach in Japan. No European advisers were recruited again until 1918, when a French military mission was invited to organize the Japanese military aviation corps.

The French had accomplished many things in nineteenth-century Japan, ranging from cadet training to artillery demonstration to development of military industries. Nevertheless, most of their pioneering work was forgotten after the arrival of the German advisers. To the Japanese, "Theory and appearance were replaced by practical knowledge and realistic performance." Among the much appreciated contributions of the German instructors was the emphasis upon efficient administration, well-trained General Staff Officers, a regularized system of inspection, divisional organization, introduction of studies in military history, preparation of national defense plans, and stress upon full-scale mobilization involving optimum use of the railroads. There can be no better testimony to the fascination that the Germans exerted upon the Japanese Army than to mention the fact that the Emperor Meiji himself came to the War College to listen to an hour-long lecture by Major Meckel—in German.

Devising a National Mission

In the Meiji era, Japan possessed no strategic war plans involving a potential enemy. The initial conception of national defense was so vague that Japan merely expected to wage a defensive campaign on her own territory if invaded. But as early as 1871 Yamagata and his associates devised a policy that foresaw the gradual shift of military emphasis from internal to external considerations. Since "the urgency of national defense against the Far Eastern policies of Tsarist Russia was felt keenly," Russia became Japan's first hypothetical national enemy. When this strong foe loomed at the "northern gate," top priority should be devoted to it. This policy became the basis for all subsequent Japanese Army planning, vis-à-vis the Asian continent, and permeated the AGS outlook throughout the seventy-five years that followed.

The first reference in modern Japanese history to the dispatch of military forces overseas occurs at the time of the first Korea crisis in 1872. This proposed expedition did not materialize, for domestic and pragmatic reasons. A minor expedition sailed to Taiwan in 1874 to crush the aborigines. The Sino-Japanese War of 1894–1895 was Japan's first great modern war, wherein she committed around 240,000 men and almost every warship. The Japanese victory within eight months was achieved at relatively slight cost, but the humiliation of the Triple Intervention led to a doubling of the number of divisions (to

NATIONAL SECURITY AND MILITARY COMMAND

thirteen). Thirty thousand Japanese troops participated in the suppression of the Boxer Uprising in 1900. Given her fine track record, Japan was welcomed into an alliance with Britain in 1902, both powers regarding Russia as the potential enemy.

The Russo-Japanese War of 1904–1905 was Japan's trial by fire. Hostilities lasted a year and a half, required commitment of 1,240,000 men (6 percent of the male population), and cost 115,000 casualties and a billion and a half yen. The equivalent of twelve provisional divisions had to be organized to reinforce the fully committed thirteen existing divisions, which were increased to nineteen regular divisions from 1907. But, more importantly, a basic national defense policy had to be devised at long last. This was accomplished concretely in 1906–1907.

The primary ground foe was judged to be Russia, which presumably would be seeking revenge for 1905. Allied to Russia was France. The other main national enemy was the United States. From this period dates the Japanese transition from a defensive to an offensive policy, now that the Russian fleet had been destroyed in Far Eastern waters, and Japan had acquired footholds on the mainland of Asia. The holdings in Manchuria were significantly enlarged after Japan annexed all of Korea in 1910.

With the exacerbation in Sino-Japanese relations following the submission of Japan's Twenty-one Demands in 1915, China became the subject of IJA operational planning contained in the annual war plans. Significantly, the Army deferred any change in national defense plans or in troop allocations documents to reflect this new anti-Chinese contingency operation, as long as a civilian was Prime Minister (Okuma Shigenobu, 1914–1916), ostensibly because of a fear of leaks. Only when General Terauchi Masakata (1916–1918) succeeded Okuma did the Army make the necessary classified changes—two years later in 1917.

Japan's participation in World War I was militarily insignificant, although the Siberian Expedition which followed proved to be a quagmire with no redeeming features. From the

Japanese standpoint, the period was characterized by remarkable changes on the international scene: the ruin of the Russian Tsardom and the birth of the Soviet Union; the enormous increase of American influence in the Far East, accompanied by a deterioration of Japanese-American relations, *e.g.*, the movement to exclude Japanese from the United States, the trade war, the naval race. As a consequence, the order of hypothetical national enemies was rearranged in 1918: #1—the United States, #2—the Soviet Union, #3—China. Japanese Army-Navy seizure of the Philippines, to deny advanced bases to the American fleet in the Western Pacific, was included in Japanese war plans for 1918.

With Russia out of action for the moment, the United States remained Japan's main problem in the early 1920s. The Japanese Army was not as serious as the Japanese Navy concerning anti-United States operations, however, because the Army did not expect hostilities in the near future. If the Philippines had to be invaded, the Japanese Army was thinking in terms of a seventy-five-day operation to preempt the arrival of the United States Navy. In the 1923 war plans, the seizure of Guam was added.

Although Japan's policy of exclusively offensive strategic operations remained inflexible thereafter, from 1928–1929 the Japanese Army began to devote serious attention again to Russia as the main enemy. A recrudescence of Russian power in the Far East would obviously endanger Japanese national aspirations on the Asian continent. It was felt more strongly than ever that Manchuria would become the battleground for the Japanese Army, as well as the front line of national defense—as far west as Lake Baikal in Siberia.

After the Manchurian Incident of 1931–1932, which carried Japan's defensive responsibilities directly to the borders of Soviet Siberia and Mongolia, the national emphasis in operational priority reverted to Russia. Admittedly, there were some, outside of IJA circles, who regarded this shift mainly as a device to obtain a larger share of the budget for the Japanese Army. The latter feared that the lion's share would go to the very expensive Navy if the United States were judged to be the primary national foe.

Japan's basic policy embodying offensive operations was still in effect when the Pacific war commenced in 1941. Actual operational planning against the British was not added until 1939; against the Dutch, until 1941. Meanwhile, of course, the Japanese Army had bogged down in China since 1937. At the outset of the "China Incident," the AGS did not contemplate committing more than eleven divisions on the mainland, or an absolute maximum of fifteen if the reserves were drawn upon. Yet, by the end of 1937, the Japanese already had to dispatch sixteen divisions and 700,000 troops, approximately the number of men in the entire standing army to date. By the end of 1939, Japanese forces in China numbered twenty-three divisions, twenty-eight brigades (the equivalent of another fourteen divisions), and an air division. The number of troops in China had risen to 850,000. This level of commitment was maintained until about 1943, when transfers to other theaters began, under American military pressure. Still, at war's end in 1945, the Japanese Expeditionary Army in China was the largest of the armies stationed overseas: 1,050,000 officers and men (19 percent of the entire IJA); twenty-six infantry divisions (15 percent of the total); plus one tank division (25 percent) and one air division (7 percent). In addition, 64,000 IJN personnel were on duty in China.

The overall figure for the size of the Japanese Army at war's end was a core of 169 infantry divisions, four tank divisions, and fifteen air divisions, supported by 9,000 aircraft. Total personnel strength was over 5,500,000 in the Army (about half in the homeland) and 1,863,000 in the Navy. This scale of military involvement is rendered all the more dramatic when we take note of the size of the Japanese armed forces in the first years of the Meiji era: in 1872 the national forces consisted of 17,096 soldiers and 2,641 sailors on active duty.

THE JAPANESE ARMY EXPERIENCE
High Command Structure

While Japan's system of command and control was excellent, indeed brilliant, during the first decades after the Meiji Restoration, the period of the next two Emperors was marked by inappropriate, incoherent, and disastrous planning and guidance. The only "legalized" high-level wartime organ of direction was the Imperial General Headquarters (IGHQ—*Daihon'ei*). Established for the first time to oversee the Sino-Japanese War and again for the Russo-Japanese War, it was not reestablished until an emergency (*jihen*) was declared in 1937. IGHQ was retained thereafter until the end of the Second World War. By regulation, the Chiefs of the Army and Navy General Staffs were to "head their respective staff offices in attending a war council called by the Sovereign, advising [him] on operations, and effecting the coordination of the Army and Navy so as to attain the ultimate purpose of the operations."

IGHQ bore only a surface resemblance to the Western type of JCS organization. Its characteristic and its weakness was that it did not function as a single joint body but as two distinct service commands. The Showa Emperor was never as martial-minded as Meiji had been, and he certainly did not attend the service conferences of the Army Section and the Navy Section, which developed their separate plans. At the joint meetings of the two sections, interservice agreements and understandings were reached, but even then separate orders went out from the Army Section and from the Navy Section. The system, it has been said, led to "duplication, oversights, and mutual recrimination," for the two services were in effect going their own ways. War guidance, *per se*, involving grand politics and strategy, was never a mandate of IGHQ, which was originally designed to cope exclusively with high command matters and was therefore staffed only by military men. With no staff officers of its own (all assigned to it served in a concurrent capacity), IGHQ really constituted, in Colonel Inada Masazumi's words, "a shadowy entity." Still, within its limitations, IGHQ functioned quite well in the area of war guidance. One of its more important roles was participation in the unofficial but useful IGHQ-Government Liaison Conference *(Daihon'ei-Seifu renraku Kaigi),* instituted since IGHQ's own reactivation in November 1937.

Logistical Considerations

The IJA attitude toward logistical factors got off to a bad start from the early days of the new national military establishment. Back around 1885, when the German military adviser Meckel was lecturing at the Army War College:

> He used the term "military command" in the broadest possible sense, stressing lines of communication and the problem of supplies for large operations. For the German word for logistics, *Etappe*, no equivalent expression could be found in Japanese; it took some effort on the interpreter's part to convey the meaning of this . . . concept. The primitive state of the supply organization may be gathered from an incident during the first staff maneuvers held under Meckel's direction. He had put two of his students (two captains) . . . in charge of the commissariat . . . . they provided a large supply of pickled plums.
>
> Meckel called the two officers "infants."

Of course, after decades of experience, the Japanese Army evolved a sophisticated logistical structure embracing transportation, communications, field ordnance, intendance, etc. But, even by war's end, the Army was chronically weak in logistical planning. One of General Yamashita Tomoyuki's headquarters officers once said: "Japanese staff officers generally find logistics work dull and thoroughly distasteful; so they skip it." Adds Colonel Hayashi Saburo: "The High Command did not awaken to the remarkable progress of material potentials in modern warfare, but instead continued to esteem the superiority of spiritual fighting strength."

THE JAPANESE ARMY EXPERIENCE
Intelligence and Operations

On the Army General Staff, the most important element, by far, was the First (Operations) Bureau. Since in other armies the intelligence (and logistics) staffs are usually accorded equal ranking with Operations, a few comments are appropriate here concerning the outlook in the Japanese Army. Under the old influence of the Prussian-German tradition, it was the IJA concept to "head straight for the sound of enemy guns." Mission, in other words, must take precedence over intelligence estimates. It became the habit for field officers to reach relatively swift combat decisions rather than to mull matters over for protracted periods. By and large, there was a tendency to treat intelligence lightly, or rather not to value it highly. Theoretically, all staff bureaus were to possess identical weight in the military decision-making process, with the Army Chief of Staff rendering the final decision, assisted by the deputy as necessary. In practice, however, priority was assigned to operations thinking, and IJA veterans generally agree that the emphasis was excessive. In General Sugiyama Gen's day, for example (the crucial period, 1940–1944), the Chief of General Staff was said to be practically a robot; real power lay with the First Bureau.

> This [Operations] bureau [comments Colonel Hayashi], with its own dictum concerning the "secrecy of supreme command," adopted ultra-secret policies and acted like a law unto itself, with utter self-satisfaction. It is the common opinion of those who served on the IGHQ staffs when the Pacific War broke out that the driving force to commence hostilities stemmed from the 1st Bureau advocates.

A complex of reasons could be adduced for this outlook—historical, strategic, demographic, and economic; but there was also the psychological factor: Deliberate or passive concepts were often deemed to be "unmilitary," whereas positive or aggressive attitudes were much esteemed and usually triumphed in the decision-making process, particularly after the outbreak of the China Incident in 1937. By extension, intelligence experts, taking the broad and sometimes pessimistic view, could be equated with anathematized passivity.

Historically speaking, Japanese intelligence had always been "in the shade" until World War I. Their work was methodical, undramatic, slow but sure. After the First World War, Japanese intelligence began to emphasize large-scale activity, not the plodding processes of old. The former reliance on detailed estimating of enemy strength began to fade away and to be replaced by what some call "risky and dangerous" intelligence work. Given the fundamental psychological outlook, inaccurate or unsound intelligence efforts, when noticed, only reinforced the extreme statements of some operations personnel: "Whatever an Intelligence staff officer could know, an Operations staff officer would *most assuredly* know!" One of my IJA interviewees went so far as to say that "Japanese Intelligence simply could not comprehend, and hence could not estimate in any way, the *scale* of modern Great Power enemies, who operate systematically, rationally, and massively."

One indictment levelled at the capabilities of IJA intelligence was that the failures stemmed from viewing matters exclusively through insular Japanese eyes: "The 'common sense' Japanese simply could not realize the 'supernatural' scale of Soviet or American thinking. Being delicate island folk, the Japanese worried about small, trivial matters, taking it for granted that the enemy would too. But he did not!" A classic example of the undervaluing of intelligence usefulness is the action taken by the Southern Army in 1942, without the approval of the high command in Tokyo, of combining the headquarters Intelligence Section with the Operations Section after the successful termination of the first phase campaigns of the Pacific war, on the assumption that the Intelligence staff had lost any *raison d'être* by then. This degradation of intelligence functions, amidst a full-scale war to the death, is underlined by the fact that the Intelligence Section of the Southern Army was not reactivated until February, 1944.

THE JAPANESE ARMY EXPERIENCE

One last reason can be adduced to explain the low estate of IJA intelligence: the Army's tendency to despise the fighting qualities of the enemy whom they had been fighting on and off, with great success, since 1894—the Chinese. One Army Chief of Staff has admitted that high command attitudes toward China were characterized by optimism and disdain in general. A section chief at headquarters in Tokyo, for instance, reacted to the news of the very dangerous armed clash at the Marco Polo Bridge at Peking in July, 1937, by telephoning his superior to say that "something interesting" had just occurred. Later a staff officer visited the general's office and berated the central staff for opposing the dispatch of troops to China. "You just do not know what the Chinese are like," complained the section chief. "That is why you are voicing such negative opinions, in anticipation of serious trouble. But, I tell you, the incident will be settled if Japanese shipping loaded with troops merely heaves to, off the Chinese coast."

This notion that China "did not count" was held widely among Japanese government circles, the general public, and the Army. One IJA commander told me how his single infantry regiment, 3,500 men strong, engaged three Chinese so-called divisions under a general officer, during the Suchow offensive in 1938, and "made mincemeat out of them." Although the casualty statistics bear him out, one has to admit that part of the Japanese military's confidence, and hence grandiose designs for defeating China, stemmed from a gross undervaluation of the Chinese Army's capabilities.

By projection, this outlook appears to have led the Japanese similarly to despise and underestimate all other possible enemies, such as Russia, Great Britain, and the United States, especially in terms of the potentialities of those powers. Observed an IJA officer after the war: "A national trait of the Japanese is that they are not interested in Intelligence. There is a Chinese proverb: 'If you know your enemy and you know yourself, you will win a hundred battles.' . . . we always fought without sufficient information."

NATIONAL SECURITY AND MILITARY COMMAND
Where Have All the Great IJA Commanders Gone?

Military histories in Western countries abound with accounts of Great Captains who were French, German, American, British, and Russian. References to Japanese chieftains are fleeting at best and, during the 1930s and 1940s, mainly concern atrocious behavior by the IJA rank and file.

Part of the reason is to be found in the fact that, after the victorious Russo-Japanese War, most of the IJA experience was acquired against the Chinese, who, as we saw, were held in low repute by the Japanese themselves, and who lacked the modern tools of war. The most recent combat experience against the Russians at Changkufeng in 1938 was small scale and at Nomonhan in 1939 was a disaster which cost commands on the IJA side; and the loss of Manchukuo in 1945 was pathetically anticlimatical. The achievements of Yamashita and, to a lesser degree, Homma Masaharu at the beginning of the Pacific war were obscured by the reverses and the humiliations of 1944–1945. A large-scale effort to invade India in 1944 led to a wretched catastrophe and vituperative recriminations that did the army commander (Mutaguchi Renya) little credit. Many otherwise able IJA commanders found little scope for skills or imagination in the constricted defense of island bastions in the western and southwestern Pacific.

In general, IJA field experience was modest in scale. Devoid of combat action of the Western Front variety in World War I, and hampered by "Have Not" realities affecting finances and resources, the Japanese Army tended to "think small' and expected to be outnumbered very seriously. Field commanders say that the Americans and Russians, for example, used artillery like machine guns, while IJA artillery employment was like raindrops by comparison. Again, a portion of the explanation for differences in outlook stems from the "atypical" nature of the warfare in China, and from the special requirements of hypothetical combat in Siberia, a region notoriously poor in

THE JAPANESE ARMY EXPERIENCE

roads and railways. Hence the Japanese Army, not unlike its French and Polish counterparts (among others) of the 1930s, was excessively "hippomobile." The Japanese automobile industry before the war was hardly in the same class as that of the Western powers, and the Kwantung Army, for one, could not comprehend how the Soviet Army could operate hundreds of miles across roadless Mongolian wastes to supply the logistical and combat needs of G.K. Zhukov's fighting army in 1939, with hundreds and thousands, not dozens, of military trucks. Again, one must remember that the Japanese homeland itself was hardly suitable for large-scale mechanized and motorized conceptions and exercises. There are very few plains, 85 percent of the of the country is not arable (that is, endowed with modest enough gradients to allow wet-rice farming), and most of the rural areas are characterized by paddy fields, narrow trails, and feeble bridges.

This is one of the unspoken reasons why the Japanese Army was so happy with its seizure of Manchuria from China in 1931–1932. Officers who served with the Kwantung Army never forgot the open spaces there. It reminded a colonel whom I interviewed of a story from the Arabian Nights, in which one could have all the land one could traverse in a day. IJA camps were now so huge that the soup was usually cold by the time it could reach the far-off mess halls. Even the blueprints for the layout of army buildings differed from those in Japan, since there was so much more space available in Manchuria.

But space alone is not all that commanders need, and the Japanese Army remained not richly endowed with armor and trucks and planes. The largest strength which the Japanese were ever able to amass in the China Theater was nine and a half infantry divisions for the drive against Hankow in 1938. Admittedly, some thought had been given, at the command level, to the possibilities of corps structure, on the basis of the experience of the European armies in World War I. The AGS operations section had devised a plan envisaging twenty army corps; but the notion was rejected as being impractical and unfeasible.

Some Conclusions

Either by preference or by necessity, the Japanese Army also reflected its *samurai* heritage by extolling flesh versus steel and the Yamato Spirit—both of which assuredly have a contribution to make, but which could by themselves hardly overcome the scientific and technological assets of twentieth-century armies. In this respect, the dogged, terribly expensive, human-bullet tactics against Heights 203 at Port Arthur in 1905 excited the imagination of the academy cadets and the staff college students. Against something like 500 to 1,000 Soviet armored cars and tanks in the Nomonhan battles of 1939, the Kwantung Army committed all of its armor, totalling eighty-five medium and light tanks. Molotov cocktails and satchel charges did not make up the difference. Hayashi agrees that the Japanese Army had not progressed beyond comprehending firepower at the levels of the Russo-Japanese War.

Apart from the tangible aspects of IJA modern experience, mention should be made of the fact that the Japanese type of general officer, ensnared by a very strict system of seniority, was in many ways a unique man. The best general was supposed to avoid involvement in detail and to adopt a noncommittal attitude, in an ancient tradition that silence was truly golden. An Olympian "general-like" attitude suggested at least superficial encouragement of each subordinate's opinion, although without commitment. Hence much of the IJA planning worked its way up rather than down, and mid-range officers *(chuken shoko)* exerted an influence on commandship which bore no correlation with their modest military ranks.

Bibliography

I have drawn upon my recording tapes of personal interviews in Japan with former IJA and IJN officers; upon the Japanese-language official war history series of the Japan Defense Agency, Office of Military History; upon documentation published by Misuzu Shobo Co.; and upon books and articles written by the Japanese authors Fukushima Shingo, Hattori Takushiro, Iwabuchi Tatsuo, Izu Kimio, Kawabe Torashiro, Kono Tsunekichi, Majima Ken, Maruyama Masao, Marsushita Yoshio, Murakami Hyoe, Ohtani Keijiro, Sumiya Mikio, Takahashi Kunitaro, Tanaka Ryukichi, Tsukuba Hisaharu, and Yamamoto Katsunosuke.

Military origins are examined in English by James B. Crowley, "From Closed Door to Empire: The Formation of the Meiji Military Establishment," in Bernard S. Silberman & H.D. Harootunian, eds., *Modern Japanese Leadership: Transition and Change* (Tucson: University of Arizona Press, 1966); Roger F. Hackett, "The Meiji Leaders and Modernization: The Case of Yamagata Aritomo," in Marius B. Jansen, ed., *Changing Japanese Attitudes Toward Modernization* (Princeton: Princeton University Press, 1965); "The Military," in Robert E. Ward & Dankwart A. Rustow, eds., *Political Modernization in Japan and Turkey* (Princeton: Princeton University Press, 1964); and *Yamagata Aritomo in the Rise of Modern Japan, 1838-1922* (Cambridge: Harvard University Press, 1971). Also see Hyman Kublin, "The 'Modern' Army of Early Meiji Japan," *Far Eastern Quarterly,* IX (Nov., 1949), 20-41; E. Herbert Norman, *Soldier and Peasant in Japan: The Origins of Conscription* (New York: Institute of Pacific Relations, 1943); Ernst L. Presseisen, *Before Aggression: Europeans Prepare the Japanese Army* (Tucson: University of Arizona Press, 1965).

For the more modern aspects, see my "Effects of Attrition on National War Effort: The Japanese Army Experience in China, 1937-1938," *Military Affairs,* XXXII (Oct., 1968), 57-62; "High Command and Field Army: The Kwantung Army and the Nomonhan Incident, 1939," *Military Affairs,* XXXIII (Oct., 1969), 302-312; and *Year of the Tiger* [Japan, domestic and external, 1937-1938] (Tokyo: Orient/West, 1964).

NATIONAL SECURITY AND MILITARY COMMAND

Specialized but important topics receive the attention of Robert K. Hall, *Shushin: The Ethics of a Defeated Nation* (New York: Teachers College, Columbia University, 1949); Yale C. Maxon, *Control of Japanese Foreign Policy: A Study of Civil-Military Rivalry, 1930–1945* (Berkeley: University of California Press, 1957). The main sources on IJA leadership to which I refer are my "Maverick General of Imperial Japan" [Lt. Gen. Sato Kotoku at Kohima, 1944], *Army*, XV, no. 12 (July, 1965), 68–75; "Qualities of Japanese Military Leadership: The Case of Suetaka Kamezo," *Journal of Asian History*, II, no. 1 (1968), 32–43; and Arthur Swinson, *Four Samurai: A Quartet of Japanese Army Commanders in the Second World War* (London: Hutchinson, 1968). For an overall survey, see Hayashi Saburo in collaboration with me, translated as *Kogun: The Japanese Army in the Pacific War* (Quantico: Marine Corps Association, 1959). Also available in English is Takata Yasuma's old but valuable *Conscription System in Japan*, ed. Ogawa Gotaro (New York: Columbia University Press, 1921).

Only a few American or British writers have dealt with Japanese commanders. I have written two articles on little-known field generals: one on Suetaka Kamezo, a division commander in the Changkufeng battle of 1938; a second on Sato Kotoku, who commanded a division in the ill-fated Kohima-Imphal campaign in 1944. John Potter, a journalist, wrote about Yamashita's career in *The Life and Death of a Japanese General*, (New York: New American Library, 1962). Others, such as Reel and Kenworthy, have concentrated on the war crimes trial period. Arthur Swinson's *Four Samurai,* despite its misleading title, is the best book-length study in English of Japanese generals in World War II. A Briton who had served in Southeast Asia, Swinson focuses on the Burma and Philippines areas; his four commanders are Mataguchi, Honda Masaki, Homma, and Yamashita.

My screening of Japanese historians' recent master lists of the most influential IJA generals shows that Yamashita's name appears frequently; Homma's occasionally; Mataguchi's rarely; and Honda's, not at all. Many Japanese authors recommend rather obvious additional names as Anami Korechika (the last war minister), Ushijima Tatsukuma (of Okinawa), Kuribayashi Tadarnichi (of Iwo Jima), and of course Tojo Hideki. But they also resurrect at least one slighted military reputation; that of Lieutenant General Miyazaki Shigesaburo, a quiet soldier's soldier. Miyazaki fought brilliantly, as a battalion commander in Jehol, as a regiment commander at Nomonhan, and as a division commander in Burma, to the very end. Western students of military commandship would do well to delve into the careers of IJA commanders, in victory and in defeat.

There is an ample repertoire from which to draw. From the Meiji period 1945, the Army had 134 full generals and 17 field marshals, while the Navy had 76 full admirals and 13 fleet admirals. At war's end, there were 21 full generals and marshals, 484 lieutenant generals, and 1,096 major generals, for a grand total of 1,601 general officers—regular, reserve, and recalled retired. As Swinson rightly says, "it is hardly sufficient to cast them merely as villains," thirty years after the Pacific War came to an end.

=== COMMENTS ===

Among the Western powers of the late nineteenth and early twentieth centuries, Great Britain and the United States escaped the full severity of the civil-military problems generated by the borrowing of Prussian-style military professionalism and the civil-military relationships that on the European continent accompanied it. Both these powers shared an English tradition of distrust of the military and careful institutional hedging-in of military prerogatives that helped to inoculate them against Prussian military autonomy and arrogance. More to the point, both were saved by geography from the necessity to maintain mass armies, except possibly in time of war. After the Prussian victories of 1870–1871, Great Britain and the United States did transform their officer corps into a military profession on the Prussian model, and they did create general staffs patterned somewhat after the Prussian example; but their histories and their insular security guarded them from the perils of a military state within the state.

Or at least that was true in peacetime. In the crisis of the Great War of 1914–1918, even the United Kingdom yielded to her military chieftains virtually autonomous control of the national capacity to make war, which in the Great War meant almost all the national capacities and national energies—until the failure of the military to repay such a concession by winning famous victories caused the civilian politicians led by David Lloyd George gradually to prize and wrestle direction of national policy back into civilian hands. And if the United States remained far enough from the cockpit of international conflict through the whole cycle of the rise and fall of Prussian-style militarism to escape ever copying Prussian civil-military arrangements, with the Second World War and the demise of Prussian militarism it can be argued with considerable cogency that the United States has developed a militarism of its own.

Ever since 1940 the United States has felt obliged to main-

COMMENTS

tain a mass army and has regarded itself as militarily insecure—two important sources of a strong military influence on the whole character of the nation-state. The American military nevertheless has not been accorded anything like the autonomy, the character of a state within the state, that distinguished the Prussian military and that the other European powers copied after 1870. But in his classic A History of Militarism, Civilian and Military *(revised edition, New York: Meridian, 1959), the German emigré Alfred Vagts recognized that militarism can be civilian as well as military—that there can occur a militarizing of the perceptions and values of civilian leaders, such that a state's policy can become militaristic while yet retaining a thorough civil control of the military. Under a leadership conditioned by a generation of military crises and by long wielding of military power so immense as to create a constant temptation to seek military solutions to otherwise intractable problems, the United States has betrayed symptoms of the influence of this civilian type of militarism. Furthermore, in the blurring of civilian and military roles, the blurring of national policy and military strategy, that developed in the United States during World War II and the Cold War, there may lie other dimensions of a newer, American-style militarism.*

The extent of the militarization of American policy since World War II is a problem to be reflected on; that an American style of militarism has arrived is not a fact about which to be dogmatic. One of the sources of concern about an American-style militarism, however, has been the rise of autonomous war-making powers in the Presidency, independent of effective Congressional restraint, in combination with a tendency for Presidents to view problems in a military perspective, tempted as they are by the great military force they command to see in that force a ready means of resolving problems. In these circumstances, the very civilian Commander in Chief who was

====== **COMMENTS** ======

established by the writers of the Constitution as one of the guarantors of civil control of the military can become a source of the militarization of American national policy.

The militarily autonomous modern Presidency first took shape during the American Civil War. For good or ill, Abraham Lincoln was the first major architect of the "imperial Presidency." Here, in a New Dimensions lecture of 1971, a historian of the Civil War reviews the evolution of the American military command system, and particularly of the President as Commander in Chief, through the Civil War. Warren W. Hassler, Jr., of the Pennsylvania State University has written among other works **George B. McClellan: Shield of the Union** *(Baton Rouge: Louisiana State University Press, 1957) and* **The President as Commander in Chief** *(Menlo Park, California: Addison-Wesley, 1971). In 1975–1976 he is the second holder of the visiting professorship in military history at Fort Leavenworth.*

High Command Problems in the American Civil War

WARREN W. HASSLER, JR.

During the American Revolutionary War, the Continental Congress acted as both executive and legislative branches in its direction of the War for Independence. This was a clumsy, awkward arrangement, and did little to assist General George Washington as American Commander in Chief of the Continental Army.[1]

There was a Board of War established in June, 1776 by the Congress, but at first it included only civilian members—a number of whom knew little of the management of war. Only later in the conflict were military men added to the board. And not until 1781 was the office of Secretary at War established, with Major General Benjamin Lincoln named to that post.[2]

The Congress and Board of War, in general, ran the hostilities in an inefficient manner. Twice, in moments of great crisis, Washington was given dictatorial power by the solons for periods of six months.[3] The Virginian set a shining example of accepting every responsibility thrust upon him by the Congress, and fulfilling them, while at the same time never forgetting the principle of civilian supremacy over the military. His model deportment and accomplishments in the face of great adversities have never been surpassed in our history. He was truly a man for all seasons during the struggle to achieve independence.[4]

George Washington also played a pivotal role in the drawing up of the Federal Constitution. In the trying months of controversy in Philadelphia during the long summer of 1787, perhaps the most difficult issue faced up to by the Founding Fathers was the complex one of the relations to be established between the executive branch as represented by the President of the United States and the legislative branch as represented by the Congress of the United States. And one of the most feared and challenging powers to be authorized was that of the command of the new nation's armed forces.[5]

What emerged from the Fathers' hands was the so-called "Commander-in-Chief Clause" of the Constitution. It states simply that "The President shall be Commander in Chief of the Army and Navy of the United States, and of the Militia of the several States, when called into the actual Service of the United States."[6] There was no elaboration. The Constitution gives Congress the right to declare war and to raise and support armies, but it provides that it is the duty of the President to wage war and to control and command the armed forces of the Republic in *both* war and peace. This arrangement drawn up by the Founding Fathers had the virtue of guaranteeing civilian control of the armed services and at the same time providing unity of command.[7]

So at the beginning of the new nation's history as an independent country, with civilian control of the military explicit and implicit in this so-called Commander-in-Chief Clause, the American polity, while supplying, when needed, a military system, was not likely to develop a militaristic system.[8] The Navy and Army were agencies of civil power, to be organized and disciplined with that idea in mind, and were not ends in themselves.

Several delegates to the Constitutional Convention contended that the President—the Commander in Chief—should not be permitted actually to direct troops in battle. But a preponderant majority of the Founding Fathers held that the Chief Executive could, if he desired, assume personal command of soldiers in the field or of warships on the water, although most of the framers thought this would hardly ever be attempted by a President.[9]

The Constitution further empowered the Chief Executive to appoint a civil and military establishment. The President reserves to himself the right to alter his orders at any time, to change commanders, or to directly interfere in any detail of command; in other words, to resume at any time the power delegated by him to his subordinates, or to control the manner of their activities. As the executive branch of the Federal Government developed, a Secretary of War (and, later, Secretaries of the Navy and Air Force and an overall Secretary of Defense) was appointed. The Secretary of War was the regular constitutional agent of the President for the administration of the Republic's military establishment, and regulations and orders publicly promulgated through him had to be received as the acts of the President and as such binding upon all within the sphere of his legal and constitutional authority.[10]

The senior general of the Army soon became known in the new nation as the Commanding General—or General in Chief—of the Army, thereby establishing, along with the General in Chief's superior, the Secretary of War, still another top subordinate to the President. It was quickly asked if these three persons could command the same force at the same time. The reply was sharp and distinct: There can be but *one* commander; all others must be commanded. To claim that the Secretary of War and/or the General in Chief command the whole Army under the President is to admit that these two subordinates do not command it! Thus, there may be any number of high-ranking officials or commanders, each dependent upon and subordinate to the next superior, and each acting upon such directives as he may receive from higher authority up to the President, but exercising supreme command within the sphere of duties left to his discretion.[11]

The constitutional conviction still remained, however, that command rather than control was exercised by the President, and this muddied the waters so far as the rapport among the Chief Executive, Secretary of War, and General in Chief was concerned. Under the Constitution, the Secretary was political, the Commanding General military, and the President both political and military. It might be normally assumed that the Secretary would be more military in function, with his obligation to represent his department's interests, than the President, whose responsibilities and interests were broader. But the Constitution reversed this relationship in part, and the hierarchy's procedure of command was somewhat obscured. For many decades it was asked whether the chain of command went up through the civilian-politician Secretary to the Chief Executive, or whether, in reality, there were not two channels of authority emanating *from* the President: a military avenue of command running straight to the highest professional officer in uniform, and a political-administrative line to the Secretary.[12]

Much of American military history has centered around this cloudy issue. The General in Chief (later the Chief of Staff) has sought to supervise both the administrative and military features of his department, as well as to secure direct access to the Commander in Chief; while the Secretary has tried to have military subordinates report directly to him while he kept exclusive access to the President. It has proved difficult throughout our history for either the Secretary or the military chief to achieve complete success in these endeavors. As Samuel P. Huntington declares: "Inevitably, the Secretary tends to get cut off from his department by a military head who oversees both military and administrative aspects, or he tends to surrender the military aspect to the professional chief who maintains a direct command relationship with the president."[13] Of course, some strong, charismatic Secretaries have been able at times to get around the horns of this dilemma.

But how would the President's power as Commander in Chief develop? Many, such as John Quincy Adams, feared the plenitude of prerogatives that the Chief Executive had as Commander in Chief. In time of war, as the Presidency has developed in the nineteenth and twentieth centuries, the Chief Executive has few curbs as to how he exercises the power of Commander in Chief, and this has held true usually even in peacetime. "Even under the stress of bitter partisanship," writes Clarence Berdahl, "and despite all its mutterings and criticisms of executive policy, Congress will be slow to deny the power of the President as Commander-in-Chief to send and maintain troops of the army and navy abroad at his discretion, or to assert any definite claim to control itself."[14]

In short, the American President occupies an almost independent position, possessing command powers unfettered by hardly any legislative or judicial restrictions. His control over civilian secretaries and military chiefs is complete. Indeed, if he did not have this power, civil control of the military would be impossible in this country!

Furthermore, specifically, the Commander in Chief has the power—despite occasional Congressional protestations—to send our armed forces anywhere, on any kind of mission. These command powers may be exercised in several areas: The Chief Executive may, of course, deploy American Armed forces in wartime as he deems best and with whatever missions he may determine, name their commanding officers, and wage the conflict as he sees fit. In peacetime, he may order warships into disputed waters, as Franklin D. Roosevelt did in the months before Pearl Harbor to perform convoy duty against German submarines, or move troops as he sees fit, as Woodrow Wilson did in 1916 when he directed General John J. Pershing into Mexico with American soldiers in hot pursuit to punish the murderous bandit, Pancho Villa. The President may also order troops into one of our states upon the request of a governor to uphold the laws and maintain public safety, as Wilson did in Colorado in 1914 during the so-called "war" between the miners and mine owners. In addition, the Commander in Chief may, upon his own volition, employ soldiers to maintain the laws of the United States, as did Grover Cleveland in Chicago in 1894 during the Pullman Strike.[15]

In any event, as Ernest R. May suggests:

> The commander-in-chief clause was almost certainly intended . . . to insure that control over the armed forces remained in politically responsible hands. . . . It was meant

to insure that America's chief executive should make those decisions that Marlborough made for [Queen] Anne. Once war has been declared, he was to determine where it was to be fought. He, though usually a civilian, was to make the choices between primary and secondary theaters. He was to assume responsibility for naming the officers to direct operations in each of these theaters and hence for their choice of subsidiary aims. He was expected, in others words, to make decisions on *priorities* and *command*.[16]

From this introductory look at the formation and nature of the Commander-in-Chief Clause, let us look now at the development, challenge, and expansion of this all-important power of the American Presidency, with special attention to the Civil War period, certainly one of the Republic's most crucial tests.

Fortunately, the United States had as its precedent-making first President the tried and experienced George Washington, a man of the highest character and wisdom, who could be relied upon to carefully develop the Commander-in-Chief power of the new Chief Executive. In his skillful handling of several expeditions against the Indians and the calling of the militia during the Whiskey Rebellion, his willingness to actually assume personal command of troops in a crisis, and his courageous stand against an unwise war with European powers, Washington performed surpassingly. His stand for an adequate peacetime military and naval establishment was also a statesmanlike action. In short, his legacy was of the highest order, and his example as the Nation's first Commander in Chief under the Constitution was invaluable.[17]

James Madison was not nearly so effective as President in the War of 1812. He was often perplexed by the complexities of the Commander-in-Chief position, and he did not act wisely in his relations with his two Secretaries of War or with the senior generals in the field. Nor did he make good choices, on the whole, of either civilian or military leaders. James K. Polk performed a bit better as a wartime President during the Mexican War of the mid-1840s, although his strong political partisanship weakened the Nation's war effort south of the border as well as north of it. Polk, too, had trouble in establishing a smoothly working system of command between himself, the Secretaries of War and the Navy, and the ranking generals in the field and in Washington. But Polk was a stronger Commander in Chief than Madison had been, and he proved conclusively, despite serious shortcomings and misadventures, that a President could "run a war" with the powers available to him.[18]

Still, relations between the Secretary of War and the General in Chief were often troubled down to the Civil War, as for example between Secretary Jefferson Davis and Commanding General Winfield Scott.[19] With hostilities upon the land, could the Union and Confederate governments effectively handle a total war of great severity in the 1860s? As will be noted, one could (after many disappointments), but the other couldn't. And the issue of personality would play a key role in the outcome.

The tall, brooding man who came out of the prairies to save the Union was little known in the North for his executive or military experience, and few expected him to handle efficaciously the onerous position of the Nation's Commander in Chief, While Abraham Lincoln had been elected a captain in the nasty little Black Hawk War of 1832, he had found it difficult to maintain discipline in his Illinois volunteer company. When some of his soldiers went off on a drinking spree, Lincoln was arrested and forced to wear for several days a large wooden sword—the badge of derision for an officer who could not handle his men in a disciplined manner. On another occasion, he fired a weapon in camp, and was again arrested. At no time had he seen action or the enemy. He later always belittled his short career in the Black Hawk War.[20]

Both the Union and Confederate governments thought the war would be a brief, limited one; instead it became a devastating total war of enormous scope, casualties, and severity. Lincoln's performance as Commander in Chief ranged from that of a well-meaning but inexperienced and often bungling experimenter in 1861 and 1862 to that of a consummate war director of high ability by 1864. He grew markedly in adeptness as well as in stature as the fratricidal conflict dragged on. The Union President became, by 1864, a good example of Clausewitz's dictum that an acquaintance with military affairs is not the sole earmark of a good civilian Commander in Chief, but that "a remarkable superior mind and strength of character" are more valuable traits.[21]

Lincoln's only organized effort to inform himself of the principles of war came when, upon his arrival in Washington, he took from the Library of Congress the book written by Henry Wager Halleck dealing with the art and science of war, and studied it. Lincoln perceived soon the impossibility of continuing limited warfare; that, to win, the North would have to completely conquer the South, not only occupying large parts of its territory and smashing its armies, but also crushing the will to resist of the people of the Confederacy. Sometimes in consultation with his Cabinet, but often without, the Chief Executive formulated policy, drew up plans of strategy, and even devised and directed tactical operations. Lincoln found it hard to communicate with many of his top generals, but he frequently tried to do this, often bypassing the Secretary of War and General in Chief. As lack of conspicuous success often attended Federal military campaigns, the President came to doubt and even to scorn the abilities of his top men in uniform. He thought he could perform better, and sometimes he could.[22]

The Commander in Chief's greatest problem was to find competent Northern generals who could perform as he wished them to. Lincoln has been somewhat unfairly criticized for appointing political generals—individuals such as Nathaniel P. Banks, John A. McClernand, and Benjamin F. Butler—and often these officers were hopelessly incompetent. But the long-range benefits of naming some political generals—often of the opposing Democratic party—did pay off in helping to set up a broad front of national purpose and unity.[23]

The President acted with admirable dispatch once secession had become a reality. With little support from the members of his Cabinet, he ordered relief expeditions to help Forts Pickens and Sumter. When the latter fort was bombarded, he called for 75,000 three-months militiamen (he shortly asked for tens of thousands more troops). He proclaimed a naval blockade of the Confederacy. He had the Secretary of the Treasury advance funds totaling two million dollars to assist in raising troops (a Congressional function). These were indeed audacious and imaginative acts for a Commander in Chief dealing with such momentous politico-military events for the first time. And the man from Springfield refused to call Congress into special session until July 4, 1861, in order to present the solons with these *faits accomplis*. When it convened, Congress rubber-stamped Lincoln's actions and tacitly approved granting him the already grasped "war powers."[24]

Furthermore, the Commander in Chief, despite Chief Justice Roger B. Taney's opinion to the contrary, suspended the writ of habeas corpus and permitted arbitrary arrests and military courts-martial, with the result that perhaps as many as 20,000 persons were imprisoned without their full constitutional rights. He went along with the forcible seizure, on occasion, of private citizens' property. He subjected passengers traveling to and from foreign nations to new passport restrictions. In addition, he early showed his generals who was master. When John C. Fremont and David Hunter, on their own authority, issued emancipation proclamations in their military departments, Lincoln promptly disavowed their actions and slapped these men down.[25]

While Lincoln was taking a firm hold on his Commander-in-Chief duties, President Jefferson Davis on the Confederate side soon showed not only certain personal limitations, but, in his relations with recalcitrant Southern members of Congress and governors, revealed the fatal flaws of a man who was Chief Executive of a would-be nation based largely on states' rights. Localism critically harmed every administrative facet of Davis' job.[26]

Probably because of his West Point training, service in the

Mexican War, and term as Secretary of War under Franklin Pierce, Davis thought himself a peerless military leader. He would have been happier, and certainly more successful, had he been merely Confederate Secretary of War. He named no less than five different men to the Secretaryship during the war. Of these individuals, only George W. Randolph and James A. Seddon could be termed competent. Leroy P. Walker and Judah P. Benjamin were, for a number of reasons, not at all suited for this high Cabinet position, and John C. Breckinridge was in office for too short a time (only several weeks) to measure his capabilities. So Davis acted as President and Secretary of War in one, and was in addition his own commanding general as well; not until February of 1865 did he name Robert E. Lee to be General in Chief of all the Confederate armies, the Confederacy having no such position until that time.[27]

Davis did not perform well in the vital area of grand strategy. Unable to delegate authority properly, he got bogged down in petty details, and he never was able to set up a regular, working Confederate command structure. As Commander in Chief of the grayclad hosts, he was largely a failure. Despite undeniable and praiseworthy talents and characteristics, he was flawed by his impatience of contradiction, his lack of tact, perception, and inner harmony, his excessive pride, and his hypersensitivity to criticism.[28]

Returning to the Federal side, for the first nine months of the war Lincoln was hampered by a Secretary of War—the elderly Pennsylvania politician Simon Cameron—who showed little initiative in or grasp of his office and duties,[29] and by a mentally sound but physically incapacitated General in Chief, the seventy-five-year-old Winfield Scott.[30] So the Commander in Chief himself took the lead in military direction. In July, 1861, despite the persistent protestations of the generals, he ordered the army of Irvin McDowell forward—to Bull Run, where it was defeated. The President then named young George B. McClellan to supersede McDowell. When McClellan and Scott failed to harmonize, Lincoln reluctantly accepted the resignation of the old "Giant of Three Wars" and named the thirty-four-year-old McClellan as General in Chief of all the Union armies. When "Little Mac" fell ill with typhoid fever, the Chief Executive—pitifully eager for an early offensive—sought, in his own words, "to borrow" the Army of the Potomac, and convened councils of war at the White House involving several Cabinet members and secondary generals. These conclaves hurried McClellan out of his sick bed.[31]

When Cameron's faulty administration of the War Department in 1861 became apparent to all, Lincoln replaced him with the dynamic though seriously limited Edwin M. Stanton, who was at least a more efficient business manager of the war office than his predecessor had been, though a petty meddler in military actions [32] (although of course, as civilian superior to all the men in uniform, it was his right to do so, if the President approved).

When McClellan was loath to launch a premature campaign in Virginia in early 1862, the Commander in Chief himself issued a series of general and special presidential war orders which dictated to McClellan not only the strategy but even the tactics he was expected to employ in his operations. The general protested in vain. Lincoln later acknowledged that these war orders were, in his own words, "all wrong." But when Little Mac finally took the field on campaign, the President fired him as General in Chief, and, along with Secretary of War Stanton, tried for some four months to handle that position himself, with little success. While no one could seriously question the Chief Executive's constitutional power as Commander in Chief so to intervene and act, the wisdom of Lincoln's conduct is open to serious challenge. He continued, however, to interfere, grossly and disastrously, in McClellan's operations.[33]

While McClellan's Peninsula Campaign was in progress in the spring of 1862, the President journeyed to Fortress Monroe, at Hampton Roads, and—so angered at alleged delays that he hurled his tall stovepipe hat on the floor—personally set in motion a small expedition across the water that helped recapture

Norfolk, which the Confederates were abandoning anyway. He then dispatched warships up the James River to succor McClellan.[34] Returning to his office in Washington, Lincoln tried tactically to maneuver troops by telegraph against Stonewall Jackson in the Shenandoah Valley, again with a conspicuous lack of success.[35] "In every country save our own," asserts the able commentator on military policy Emory Upton, "the inability of unprofessional men to command armies would be accepted as a self-evident proposition."[36]

But in the Western Theater of Operations, where the Commander in Chief had such generals in early high position as U.S. Grant, William T. Sherman, George H. Thomas, and William S. Rosecrans, victories of an almost unbroken nature were achieved. Lincoln kept his hands off, largely, of the details of these Mississippi Valley campaigns, and he backed these Western generals strongly.[37] Unfortunately for the Union cause, he failed to give similar support and a hands-off attitude to Eastern generals of ability, such as McClellan and George G. Meade.

Commencing in late 1861, the Commander in Chief—and the more conservative Democratic generals—were terribly hamstrung when Congress, determined to insinuate itself into the direction of the war effort, established the notorious Joint Committee on the Conduct of the War. This nefarious group—dominated by Radical Republicans—called generals before it, put them under oath, and encouraged them to betray campaign secrets and to criticize brother officers of a more conservative persuasion. The Committee even tried to bully-rag Lincoln and browbeat him into more hasty military action. But the Committee was only partly successful in this latter endeavor, as Lincoln was determined to remain as independent a Commander in Chief as possible.[38]

The President, however, demonstrated that he still had much to learn when, in the summer of 1862 he appointed the largely incompetent Halleck as General in Chief. He then supported the latter's insistence on withdrawing McClellan's army from the vicinity of Richmond to the Washington front. At worst, McClellan had had a drawn campaign up to that time. This blunder was compounded when Lincoln and Halleck named the hopelessly incapable John Pope to command the main Federal Army in the East. Pope was promptly thrashed by Lee at Second Manassas in late August, 1862.[39]

Lincoln was confronted with a grave crisis when Lee followed up his victory by an invasion of the North. The Union President responded with a touch of the inspiration of a great Commander in Chief—as he would show a number of times later in the war—when he courageously reappointed McClellan to the command of the Army of the Potomac despite nearly unanimous opposition from members of his own Cabinet.[40] McClellan justified the trust of the Chief Executive by at least gaining a strategic victory at the Antietam in September, 1862, forcing Lee to abandon the battlefield—including many Confederate dead and wounded—give up his invasion of Union territory, and retreat back into Virginia.[41]

However, when McClellan stated that he was unable to pursue Lee at once, serious doubts arose again in the President's mind. In addition to questions as to McClellan's intentions and abilities, there were the fall elections approaching. Although a more patriotic Unionist than McClellan never breathed air, Lincoln thought—unfairly, in this case,—that the general was playing false. So the Commander in Chief removed Little Mac from command in early November and replaced him with the befuddled Ambrose E. Burnside, who had not even been a reliable corps commander.[42] Assuredly, the President had good intentions and felt he was acting in the best interests of the Union war effort, but he was far from being a good judge of military men in the first half of the war. Burnside at first procrastinated more than his predecessor had done, and then surged forward to a catastrophic defeat at Fredericksburg in mid-December. As a result, desertions in the Army of the Potomac reached an all-time high in January, 1863. Once again the bereaved Chief Executive had to remove his army commander and seek another.[43]

NATIONAL SECURITY AND MILITARY COMMAND

The mantle of command fell in late January upon "Fighting Joe" Hooker, an excellent corps commander, but one who had a hinged tongue. Not only did Hooker denounce some other officers—including Burnside and McClellan—but he demonstrated he could play the political game by snuggling up to the Radicals in Congress, who favored him as one of their own.[44] In making the appointment, the Commander in Chief wrote Fighting Joe an astonishing letter. It read:

> I think it best for you to know that there are some things in regard to which I am not quite satisfied with you. . . . I think that during General Burnside's command of the army, you have taken counsel of your ambition, and thwarted him as much as you could, in which you did a great wrong to the country and to a most meritorious and honorable officer. I have heard, in such a way as to believe it, of your recently saying that both the army and the Government needed a dictator. Of course, it was not for this, but in spite of it, that I have given you the command. Only those generals who gain successes can set up dictators. What I now ask of you is military success, and I will risk the dictatorship.[45]

As the historian Edward Channing said of this missive, "It was an extraordinary letter to write to one whom a great place had just been given, and seems to carry in itself conclusive reasons why the appointment should not have been made."[46] "He talks to me like a father," said a temporarily chastened Hooker with emotion. "I shall not answer this letter until I have won him a great victory."[47]

But at Chancellorsville in early May, 1863, Hooker sustained, instead, a great defeat. Lee followed up his advantage the next month by embarking on his second invasion of the North, thrusting irresistibly into the heart of southern Pennsylvania. Apparently having decided that the forthcoming critical battle should not be fought by Hooker, the President, Stanton, and Halleck intervened in the tactics of the campaign in its first stages to such a degree that Fighting Joe submitted his resignation, which was promptly accepted.[48] Then, in the crucial hour, Lincoln chose solid, reliable, unspectacular George Gordon Meade to command the Army of the Potomac, and gave him almost a free hand. The Chief Executive thought that the Pennsylvania general would "fight well on his own dung-hill."[49]

As the historian J.G. Randall simply expresses it, "Accepting extraordinary prerogatives conferred by the President, [Meade] prepared to fulfill a responsibility unexcelled, unless by Washington, in previous American history."[50] He met the crisis by masterfully winning the pivotal victory at Gettysburg in early July, and forcing Lee to fall back into Virginia. But when the gray chieftain retired safely across the Potomac, the patience of the sorely-tried President snapped and he lashed out at Meade for having "permitted" Lee to do this.[51] In high dudgeon, Meade tendered his resignation, feeling—correctly—that the sharp criticism was unwarranted. But Lincoln swiftly regained his magnificent equipoise, turned down the resignation, and retained Meade in command for the time being.[52]

In early 1864, Congress recreated the grade of lieutenant general (which rank no officer had held since George Washington). Lincoln appointed U.S. Grant to that grade and also named him General in Chief to succeed the ineffective Halleck. The latter, in this streamlining of the Union high command, was made chief of staff to Grant at Army headquarters in Washington, where he was of value as a communications link between the civilian Commander in Chief and Secretary of War on the one hand and Grant and the top theater and army generals on the other. In short, the President had finally set up a modern command system, and it was one of his most noteworthy achievements.[53]

Unlike the first eighteen months of the war, when for example, Lincoln had insisted on knowing the details of most of McClellan's plans and had even thrust some minute tactical ones of his own upon that general, the President in early 1864 gave Grant considerable autonomy and latitude. While Meade officially remained in command of the Army of the Potomac, Grant established his General in Chief's headquarters with that army, and in effect commanded it, Meade serving as little more than an executive officer.[54] "The President told me," reported Grant, with a little exaggeration, "[that] he did not want to know what I proposed to do.... I did not communicate my plans to the President nor did I to the Secretary of War or to General Halleck."[55]

However, while the Commander in Chief thus allowed Grant, as ranking general, wide discretion of action, he never failed to keep well informed of things military during the last fourteen months of the war, and he always made sure that Grant *knew* that he, the President, was ever at the apex of the Federal military structure.[56] When, in early 1865, Grant showed intentions of meeting with General Lee to talk about affairs above and beyond the capitulation of the Confederate Army, the Chief Executive brought him up sharply. Grant received the following telegram from Secretary Stanton, drafted by Lincoln himself: "The President directs me to say to you that he wishes you to have no conference with General Lee unless it be for the capitulation of General Lee's army, or on some minor and purely military manner. He instructs me to say that you are not to decide, discuss, or confer upon any political question. Such questions the President holds in his own hands; and will submit them to no military conferences or conventions."[57]

With this efficacious Union military high command performing creditably, and with Southern forces being depleted by attrition to the point where few reinforcements were available, Lee and Joseph E. Johnston surrendered the two main grayclad armies in April, 1865, to Grant and Sherman, respectively. In the light of the Commander in Chief's brusque reminder to Grant about involving himself in political affairs, it is of considerable interest that Lincoln now approved certain parts of the peace terms which Lee readily accepted from Grant and upon which basis he surrendered his Army of Northern Virginia. In this matter Grant certainly exceeded his sphere of authority. Thus, considerable significance is attached to the following italicized words in the historic surrender document, especially in view of the bitter policies of the Congressional Radicals during the ensuing Reconstruction years:

> ... The [Confederate] officers to give their individual paroles not to take up arms against the Government of the United States until properly exchanged, and each company or regimental commander [to] sign a like parole for the men of their commands. .... This done, each officer and man will be allowed to return to their homes, *not to be disturbed by United States authority so long as they observe their paroles and the laws in force where they may reside.*[58]

Actually, neither Grant nor even the President had the authority to bind over succeeding administrations, officers, or Congresses to refrain from "disturbing" those who had risen in rebellion against the Federal Government and Constitution—especially high-ranking Confederate civil officials and military officers. Yet the Chief Executive's magnanimity overcame his insistence upon civilian supremacy to the extent that he accepted this unusual proffer which Grant extended to Lee and the Confederates.[59] "With malice toward none; with charity for all."

By the time of the final Confederate surrenders in the spring of 1865, Abraham Lincoln had, in this vast and increasingly skillful expansion of the role of the President as Commander in Chief, bequeathed a heritage to successors that would be hard to equal, let alone surpass.[60] Well might Stanton say, upon the assassination of the man from the prairies, "Now he belongs to the ages."[61]

Notes

1. C. Joseph Bernardo and Eugene H. Bacon, *American Military Policy: Its Development Since 1775* (Harrisburg: Stackpole, 1955), pp. 22–25.

2. Chauncey W. Ford, ed., *The Journals of the Continental Congress, 1774-1789* (Washington: Government Printing Office, 1903), V: 434–435; VII: 216; IX: 818–819.

3. Emory Upton, *The Military Policy of the United States* (Washington: Government Printing Office, 1912), pp. 23,30.

4. See Douglas Southall Freeman, *George Washington: A Biography* 7 volumes, (New York: Scribner, 1948-1957), III, IV, V.

5. Warren W. Hassler, Jr., *The President as Commander in Chief* (Menlo Park, Calif.: Addison-Wesley, 1971), p. 6.

6. The Constitution of the United States, Article II, Section 2.

7. Louis Smith, *American Democracy and Military Power* ... (Chicago: University of Chicago Press, 1951), pp. 25-26.

8. Alfred Vagts, *A History of Militarism: Civilian and Military* revised ed., (New York: Meridian, 1959), p. 103.

9. Samuel P. Huntington, *The Soldier and the State: The Theory and Politics of Civil-Military Relations* (Cambridge: Harvard University Press, 1957), pp. 184-185.

10. James B. Fry, *Military Miscellanies* (New York: Brentano's, 1889), pp. 90, 97.

11. *Ibid.*, p. 93; Hassler, *The President as Commander in Chief*, pp. 9-10.

12. Huntington, *The Soldier and the State*, p. 186.

13. *Ibid.*, pp. 184, 186-189.

14. Clarence A. Berdahl, *War Powers of the Executive in the United States* (Urbana: University of Illinois Press, 1921), pp. 124-125.

15. Hassler, *The President as Commander in Chief*, p. 13.

16. Ernest R. May, *The Ultimate Decision: The President as Commander in Chief* (New York: Braziller, 1960), p. 19.

17. Freeman, *George Washington* VI, VII, *passim*.

18. See Hassler, *The President as Commander in Chief,* chaps. III, IV.

19. Charles Winslow Elliott, Winfield Scott, *The Soldier and the Man* (New York: Macmillan, 1937), pp. 386-387, 574, 649-650, 653-660.

20. Benjamin P. Thomas, *Abraham Lincoln: A Biography* (New York: Knopf, 1952), pp. 31–33.

21. Karl von Clausewitz, *On War,* trans. O.J. Matthijs Jolles (New York: Random House, 1943), p. 599.

22. T. Harry Williams, *Lincoln and His Generals* (New York: Knopf, 1952), pp. 7-14.

23. Fred Harvey Harrington, *Fighting Politician: Major General N.P. Banks* (Philadelphia: University of Pennsylvania Press, 1948), pp. 54-56.

24. J.G. Randall and David Donald, *The Civil War and Reconstruction* (Boston: Heath, 1961), pp. 168, 177-178, 274-276.

25. *Ibid.*, pp. 293-309, 371-372.

26. Clement Eaton, *A History of the Southern Confederacy* (New York: Macmillan, 1954), pp. 46-51, 263-269.

27. Frank E. Vandiver, *Rebel Brass: The Confederate Command System* (Baton Rouge: Louisiana State University Press, 1956), pp. 23-63.

28. Rembert W. Patrick, *Jefferson Davis and His Cabinet* (Baton Rouge: Louisiana State University Press, 1944), *passim*.

29. See A. Howard Meneely, *The War Department, 1861: A Study in Mobilization and Administration* (New York: Columbia University Press, 1928), *passim*.

30. William H. Russell, *My Diary, North and South* (Boston: T.O.H.P. Burnham, 1863), p. 148.

31. Warren W. Hassler, Jr., *Commanders of the Army of the Potomac* (Baton Rouge: Louisiana State University Press, 1962), pp. 10-24, 34.

32. A.K. McClure, *Abraham Lincoln and Men of War-Times ...* (Philadelphia: Times Publishing Co., 1892), pp. 155-162; John T. Morse, ed., *The Diary of Gideon Welles ...* 3 vols., (Boston: Houghton Mifflin, 1911), I: 55–69, 127–129, 148–149, 203, 234; Hugh McCulloch, *Men and Measures of Half a Century ...* (New York: Scribner, 1889), p. 301; Theodore C. Pease and J.G. Randall, eds., *The Diary of Orville Hickman Browning . . .* 2 vols., (Springfield: Illinois State Historical Library, 1925-1933), I: 533, 538–539; Noah Brooks, *Washington in Lincoln's Time* (New York: Century, 1895), pp. 28-29; U.S. Grant, *Personal Memoirs of U.S. Grant* 2 vols., (New York: Charles L.

Webster, 1886), II: 105, 123, 506, 536, 537; Charles A. Dana, *Lincoln and His Cabinet* (Cleveland: De Vinne Press, 1896), pp. 20, 26–27; Howard K. Beale, ed., *The Diary of Edward Bates, 1859–1866* (Washington: Government Printing Office, 1933), pp. 228, 381, 391.

33. Warren W. Hassler, Jr., *General George B. McClellan: Shield of the Union* (Baton Rouge: Louisiana State University Press, 1957), pp. 52ff.

34. Joseph B. Carr, "Operations of 1861 About Fort Monroe," Robert Underwood Johnson and Clarence Clough Buel, eds., *Battles and Leaders of the Civil War* 4 vols., (New York: Century, 1884–1888), II: 152; cited hereafter as *B. & L.*

35. Colin R. Ballard, *The Military Genius of Abraham Lincoln* (Cleveland: World, 1952), chap. VIII.

36. Upton, *Military Policy of the United States*, pp. 323, 394–395.

37. See John Fiske, *The Mississippi Valley in the Civil War* (Boston: Houghton Mifflin, 1900), *passim*; Bruce Catton, *Grant Moves South* (Boston: Little, Brown, 1960), *passim*.

38. *Committee on the Conduct of the War: Reports, 1863*, I: 280; 1865, Supplement, II: 110–111, 189–190, cited hereafter as *C.C.W.*; William W. Pierson, Jr., "The Committee on the Conduct of the Civil War," *American Historical Review*, XXIII (April, 1918), 550–576.

39. Hassler, *The President as Commander in Chief*, p. 64.

40. *The War of the Rebellion: A Compilation of the Official Records of the Union and Confederate Armies* (Washington: Government Printing Office, 1880–1901), XI, Pt. I, 105; George B. McClellan, *McClellan's Own Story: The War for the Union . . .* (New York: Charles L. Webster, 1887), pp. 535, 545, 566; *B. & L.*, II: 549–550. Morse, ed. *Diary of Gideon Welles*, I; 104–106; *Diary and Correspondence of Salmon P. Chase* (Washington: Government Printing Office, 1902), pp. 64–65.

41. Francis W. Palfrey, *The Antietam and Fredericksburg* (New York: Scribner, 1882), chap. III.

42. Hassler, *General George B. McClellan*, pp. 297–329.

43. Hassler, *Commanders of the Army of the Potomac*, pp. 101–123.

44. Walter H. Hebert, *Fighting Joe Hooker* (Indianapolis: Bobbs-Merrill, 1944) *passim*.

45. *O.R.*, Serial No. 40, p. 4.

46. Edward Channing, *A History of the United States* 6 vols., (New York: Macmillan, 1905-1925), VI: 477.

47. John G. Nicolay and John Hay, *Abraham Lincoln: A History* 10 vols., (New York: Century, 1890), VII: 87-88.

48. *C.C.W.*, 1865, I: 150-151, 173, 176-178, 290-291, 292; *O.R.*, Serial No. 43, pp. 58-60; *B. & L.*, III: 270.

49. George C. Gorham, *The Life and Public Services of Edwin M. Stanton* 2 vols., (Boston: Houghton Mifflin, 1899), II: 98-100.

50. J.G. Randall, *Lincoln The President: Springfield to Gettysburg* 2 vols., (New York: Dodd, Mead, 1945), II: 275.

51. Allen Thorndike Rice, ed., *Reminiscences of Abraham Lincoln by Distinguished Men of His Time* (New York: North American Publishing Co., 1886), p. 402; David Homer Bates, *Lincoln in the Telegraph Office ...* (New York: Century, 1907), p. 156; McClure, *Lincoln and Men of War-Times,* p. 360; Morse, ed., *Diary of Gideon Welles,* I; 370; Nicolay and Hay, *Abraham Lincoln,* VII: 278.

52. *O.R.* Serial No. 43, pp. 92-94; George G. Meade, *Life and Letters of George Gordon Meade* 2 vols., (New York: Scribner, 1913), II: 134, 135.

53. T. Harry Williams, *Americans at War: The Development of the American Military System* (Baton Rouge: Louisiana State University Press, 1960), pp. 77-79.

54. George R. Agassiz, ed., *Meade's Headquarters, 1863-65; Letters of Colonel Theodore Lyman* (Boston: Little, Brown, 1922), p. 224.

55. Grant, *Personal Memoirs,* II: 122-123.

56. Williams, *Lincoln and His Generals,* pp. 304-306, 349-350.

57. Stanton to Grant, March 3, 5, 1865, Roy P. Basler, ed., *The Collected Works of Abraham Lincoln* 9 vols., (New Brunswick, N.J.: Rutgers University Press, 1953), VIII: 330-331; *O.R.,* XLVI: Pt. II: 802, 841.

58. *B. & L.,* IV: 739.

59. See Nicolay and Hay, *Abraham Lincoln,* X: 196.

60. Frederick Maurice, *Statesmen and Soldiers of the Civil War: A Study of the Conduct of War* (Boston: Little, Brown, 1926), pp. 114, 116, 121, 153.

61. Nicolay and Hay, *Abraham Lincoln,* X; 302.

========== COMMENTS ==========

For a century, from the end of the War of 1812 to the intervention of the United States in the First World War, the American Army did not have to fight the army of any of the great powers. Modern American military history—in its operational aspects and in its civil-military complexities—begins principally with the First World War, the entry of the United States into the military major leagues. For American military commanders, the transition from leadership of what had become mainly a constabulary force to command of forces that must confront the armies of Germany was as formidable a challenge as any soldiers of a great power are likely to face. They had begun preparing themselves, organizationally and intellectually at least, with the Elihu Root reforms of the opening of the twentieth century, which created the General Staff and rounded out the Army school system with the Army War College. But on April 6, 1917, the American Army was still far from approximating the armies of the European great powers on almost any level—except, fortunately, in the professional competence of its best officers. How those best officers met the challenge of an abrupt shift from a constabulary army to a great-power army was sketched in a 1973 lecture by Edward M. Coffman of the University of Wisconsin. At work, when he gave the lecture, on a book that will fill in much of the larger background of the subject—a social history of the Army during its transition from old to new Army —he had already published two books on the American war effort in 1917–1918, The Hilt of the Sword: The Career of Peyton C. March *(Madison: University of Wisconsin Press, 1966) and* The War to End All Wars: The American Military Experience in World War I *(New York: Oxford University Press, 1968).*

American Command and Commanders in World War I

EDWARD M. COFFMAN

SEPTEMBER 13 is a most appropriate day to talk about American commanders in the First World War. It was the official birthday of the one American general of that war of whom most people have heard—John J. Pershing. The terms "was" and "official" may be useful because Pershing's most recent biographer, Donald Smythe, has discovered that the date is false. Apparently the future General of the Armies lied about his age, changing it probably from January 13 to September 13 in order to be within the maximum age limit for entering cadets at the Military Academy.

The American Regular Army numbered only some 127,500 (including 5,791 officers) at the time the United States intervened in World War I in April 1917. It was a small army in comparison with the large armies of the belligerents, and it had neither conscription nor a comprehensive reserve system to enable it to expand rapidly in an emergency. At that, this 1917 army was five times larger than the tiny 25,000-man army that virtually all of the senior American officers of World War I had entered as lieutenants during the last quarter of the nineteenth century.

The Civil War dominated the thought and the life of the Army throughout the early careers of the World War I commanders. Those who attended West Point in the seventies and eighties were personally exposed to the great names. The superintendents from 1876 to 1889 were famed Union commanders—John M. Schofield, Oliver O. Howard, Wesley Merritt, and John G. Parke—and the visitors were even more impressive.

George B. Duncan, a member of Pershing's Class of 1886 and a division commander in 1918, recalled:

> We often saw at West Point Generals Grant, Sherman and Sheridan, back to visit friends and to imbibe the atmosphere of their Alma Mater. To see these great soldiers was an inspiration to all of us. They were all present at the unveiling of the Thayer Monument [during Duncan's plebe year in the spring of 1883], seated side by side. After the ceremonies, the crowd formed in line to shake hands with General Grant; he just dropped his hand over the side of the rail, and continued his smoke and conversation with the other two generals. The procession came by, shaking the inert hand of a dangling arm. The General paid not the slightest attention to any individual who shook that hand.

Among the seventy-seven graduates in this class were, in addition the General of the Armies, fifteen generals and ten brigadier generals of World War I vintage. Smythe also points out that more than a quarter of the World War I generals were at West Point during Pershing's years there.

During their first-class year, Grant died and Pershing as First Captain commanded the Corps of Cadets when it presented arms at the station as the funeral train passed through. He recalled this as "the greatest thrill of my life." Then and later, he regarded Grant "as the greatest general our country has produced."

The Army was so small that there were not enough vacancies for all of the seventy-seven of the Class of '86. Congress, however, did give a special dispensation and commissioned those without vacancies as additional second lieutenants to be promoted to second lieutenant whenever vacancies occurred. What they found when they reported to their first posts might well have caused doubts as to their wisdom in choosing a military career. Duncan was one of those:

Fort Wingate, New Mexico, when I joined the 9th Infantry was the home station of several troops of the 6th Cavalry and 13th Infantry; other companies were in camp waiting assignments to permanent stations. All had just been paid. The post trader's store was in full blast. The part reserved as the officers' club was filled night and day and there were card games galore with stakes from drinks to twenty dollar gold pieces. The bar for enlisted men was lined up until taps. Intoxication was evident on all sides. Of course there were exceptions but with all there was tolerant acceptance of conditions. No military duties were attempted beyond guard mounting and required roll calls.

To my unsophisticated mind this introduction to an army post made a deeply unfavorable impression and a regret that I had not resigned after graduation and taken a job which had been offered me on the New York Central Railroad. After five or six days the Commanding Officer called a halt by placing some officers in arrest and threatening court-martial.

The Army was stagnant. Civil War veterans—overage in grade and who had often held much higher rank in their youth during the war—held virtually all positions of responsibility. In 1888, when Peyton March joined the Third Artillery at Washington Barracks, he had a Mexican War veteran as his regimental commander and more than half of his fellow officers were Civil War veterans. As late as 1900, all regimental commanders in the infantry, cavalry, and artillery and some captains were veterans of that war. Old men dominated the Army down to and including company level.

Then too, there simply was not much to do. T. Bentley Mott, another of Pershing's classmates, remembered that during his first two years in an artillery battery in the garrisons around San Francisco Bay—"our whole military work consisted of dress parade and one hour of drill each day. . . ." Another artillery officer, who was commissioned from the ranks in 1898,

Richard McMaster, recalled that one officer of this era when asked what his duties were answered "mostly social." McMaster added: " . . . that was about it. As a result many took up fads. I remember one who wove rugs—another carved wooden figures and so on." Mott summed up life in this period: "Nobody who has not seen it can picture the desolate narrowness of life in army garrisons during the 'eighties and 'nineties, or imagine the intellectual rigidity of almost all the men who, with an iron hand, commanded them."

The Spanish War, followed closely by the Philippine Insurrection, livened things up. Congress increased the strength in the Regular Army, and many of the officers and men had to act as governmental administrators as well as combat leaders. And in Mindanao there would be active service virtually up to World War I. William H. Simpson, who commanded the Ninth Army in World War II, recalled recently that he spent three-fourths of his tour in Company E, 6th Infantry, at Camp Keithley on Lake Lanao, 1910–1911, in the field chasing Moro outlaws. And General Pershing fought a full scale battle against the Moros at Bud Bagsak on Jolo in June, 1913. In reminiscing about the Army he served in from 1909 to 1917, General Simpson pointed out that it was "almost a Civil War Army." Infantry relied on rifles and pistols. As he recalled, there were only two machine guns in his regiment (Benet-Mercier), and neither would operate. In those days, "the soldiering was kind of simple. . . ."

One of the most striking things about garrison life of this era is that officers did not seem to have much to do; indeed, neither did enlisted men. Close-order drill, perhaps a class or two, or a tactical exercise on the parade ground—not in the field—in the morning; then the rest of the day was free. Experienced noncoms were expected to handle routine problems. Officers retired to the club for lunch, followed by an afternoon of bridge or perhaps polo if in the cavalry or artillery.

Mott had pursued an interesting career as a military attaché in Europe, but in 1913 he went to Camp Stotsenberg in the Philippines as a lieutenant colonel in a field artillery regiment. "I had absolutely no work to do from morning until night. . . ." The regimental commander sympathized with him about this predicament; but "as he put it, he couldn't give me anything to do without taking it away from the battalion commander or himself, and he had promised the job of building a golf course to an 'extra' cavalry colonel having no more to do than myself." Faced with this dull life, Mott retired—only to return to active duty during the war when he served as Pershing's liaison with Marshal Ferdinand Foch. It was, as William A. Taft's Secretary of War, Henry L. Stimson, put it, "a profoundly peaceful army."

There were some intellectual stirrings. At Fort Wright, Washington, a battalion post of the 14th Infantry in 1915–1916, Omar Bradley, just out of West Point, joined a group of other second lieutenants, led by Forrest Harding, who met in each other's quarters to discuss tactical problems. Incidentally, Harding, a West Point graduate of 1909, did not make first lieutenant until July, 1916. And this was not unusual. A check of his class in G.W. Cullum's *Register* shows that while coast artillerymen and engineers were promoted in two or three years respectively, field artillery, cavalry, and infantry officers had to wait six or seven years. The latter group included, George S. Patton, who made first lieutenant on May 23, 1916; Jacob L. Devers, April 1, 1916; Robert L. Eichelberger, July 22, 1916; and William Hood Simpson, July 1, 1916.

There were also the schools where the Army attempted to formalize the intellectual proclivities of selected officers. At Forts Leavenworth, Monroe, and Riley and at the Army War College, these officers devoted full time to study of the military art. Leavenworth, in particular, had undergone an exciting renaissance after the turn of the century. The young officers— George C. Marshall, Hugh Drum, and others—who occupied the key staff positions in the AEF profited greatly from this training.

Apparently there was little thought that the United States might intervene when war broke out in Europe in 1914. The

European military experts assumed that it would be a short war anyway. As it progressed without any decision, military and some civilian groups began to support preparedness. These advocates did not propose intervention but rather defense in case of invasion. Pervading the movement was the belief that military training would make a better and healthier citizen of the average American youth.

War, if it came, seemed more likely to be with Mexico, where political upheaval had brought about almost continuous friction since 1911. In the spring of 1914, President Wilson sent an expedition into Veracruz and kept it there for several months. Again, in March, 1916 he dispatched a force (consisting of four cavalry and two infantry regiments and two batteries of field artillery initially) into northern Mexico and followed this up by mobilizing the National Guard and ordering it to the border. These operations proved to be valuable rehearsals. John J. Pershing, the commander of the Punitive Expedition, was thus tested by his superiors. He, in turn, tested his subordinates and was able to form opinions as to the fitness for command and staff assignments of these men. This was in addition to the value of the field service and training for both Regulars and Guardsmen.

The German gamble on unrestricted submarine warfare in the winter of 1916-1917 precipitated the chain of events leading to President Wilson's appeal to Congress for a declaration of war on April 2, 1917. At that time, neither Wilson, who was never very much interested in the military, nor Congress fully realized what this meant. The general assumption was that the United States would extend financial and material aid to the Allies and perhaps some naval reinforcement. Within weeks, visiting Allied missions spelled out the situation and made their wants known. While they did not at that time press for men, they indicated that they would welcome a "show-the-flag" token force.

The government began to mobilize the economy and to expand the armed forces. Congress enacted conscription, and workers hastily built camps for the hundreds of thousands of draftees. The Army gave priority to officers' training camps, which turned out their first ninety-day graduates in the summer of 1917.

At the senior level, Secretary of War Newton D. Baker had to decide which officers he would recommend for stars. He described his method in a letter to President Wilson on May 30, 1917, in which he sent the names of three brigadier generals and nineteen colonels to be promoted. Secretary Baker pointed out that customarily the War Department apportioned the brigadier-general appointments to the various arms. Then he continued:

> I began by taking a confidential vote among the general officers of the Army and after tabulating this vote went over the record of each man with a view to eliminating any whose records seemed to indicate they ought to be passed over. I then made my final selections on the basis of a conference with General [Tasker H.] Bliss, General [Ernest A.] Garlington and General [Henry P.] McCain [these were the Acting Chief of Staff, Inspector General, and the Adjutant General respectively] and this list is the outcome.

Baker concluded that although he had not adhered strictly to seniority, he had taken it in consideration, but he had emphasized the mental and physical vigor of the officers in question and the opinions of their superiors. Hugh L. Scott, the Chief of Staff at this time, wrote another major general, J. Franklin Bell, on September 5, 1916 in regard to an earlier promotion list. He said then that "in most cases" Baker went along with the majority vote of the generals.

A great personality threatened to disrupt the process of selection of commanders early in the war. Theodore Roosevelt had been planning for some time a Rough Rider-like volunteer division with himself as commander and outstanding Regular officers in key positions. He proposed this to the Secretary of War shortly after the United States entered the war. On the strong advice of the Chief of Staff, Hugh Scott, who argued that granting this request would thwart plans for the systematic raising and officering of the Army, Baker and Wilson flatly rejected the former President's urgent pleas. This meant that unlike the past, there would be no political generals.

Baker's most critical decision as to commanders came when he chose John J. Pershing as the commanding general of the expedition to France. Pershing's handling of the Punitive Expedition made him the obvious choice for this assignment. No other officer had had the opportunity to demonstrate his abilities in such a command during Baker's tenure in the War Department. Once he selected Pershing, Baker backed him completely.

As Commander in Chief of the American Expeditionary Force, Pershing occupied a semi-autonomous position. This led to subsequent problems in relations with the War Department, since Pershing considered his requests as orders. As long as the Chief of Staff subordinated himself to Pershing this system worked after a fashion, although at times Pershing's lack of knowledge and understanding of the home front situation created difficulties. Tasker H. Bliss, who actually carried the load as Chief of Staff throughout most of the first seven months of the war, later said that he considered his position as Chief of Staff in reality to be that of Assistant Chief of Staff to the Chief of Staff of the AEF. When a strong man—Peyton C. March, came to the office in the spring of 1918—became Chief of Staff of the Army, friction developed between the War Department and Pershing's headquarters.

Pershing was a hard, tough soldier whose service in Mindanao had won him a jump from captain to brigadier general during the Roosevelt administration. General Ben Lear, who was a lieutenant in Pershing's troop in the 15th Cavalry during the Moro War, told a story that reveals one side of the Pershing personality.

> One day I was in the outpost area and was fired on by one of our outposts. I happened to mention it to the adjutant. I was not complaining. Pershing who overheard the

conversation said: "He should have killed you." I thrust out my chest . . . and said: "I disagree with you, sir." Wasn't that a harsh thing to say to a kid lieutenant? I never heard him praise anyone. The Army was harsher in those days than it was in World War II.

General of the Army Omar N. Bradley confirmed Lear's comparison of the World War I and II approach to leadership. He recalled an incident that took place when he was with the Veterans Administration after the war. As he was returning from Atlanta, he filled up his plane with servicemen hitching rides to Washington. As soon as the plane took off, soldiers crowded around him, some talked about the war, and others asked him for his autograph. After a while they settled down and one of Bradley's staff remarked, "times had certainly changed." This veteran of World War I remembered on one occasion that he had walked a mile out of his way to avoid General Pershing. Bradley concluded that he didn't know whether or not his and Eisenhower's approach was better than Pershing's. Of course the men were different from their fathers and expected a different relationship with their officers. Ironically, Ben Lear, who had the reputation of being a martinet, found that out in the summer of 1941 when he disciplined some Guardsmen for yelling at girls on a golf course. The resulting furor in the press indicated the great change since 1918.

There was more to Pershing than this hardness. He had a phenomenal memory. His classmate Duncan remarked on another valuable trait, that he "had the quickest eye and most accurate estimate of a body of troops as to their defects or to their efficiency of any officer that I have ever known." He also had presence, which served him in good stead when he went to Europe. There, as a mere major general without experience in this war, he had to deal on equal terms with field marshals and premiers—all of whom were very experienced. James G. Harbord commented on this during their first month in France: "General Pershing certainly looks his part since he came here. He is a fine figure of a man; carries himself well, holds himself on every occasion with proper dignity; is easy in manner, knows how to enter a crowded room, and is fast developing into a world figure." For his great work in the AEF, Pershing relied heavily upon his ability to pick able subordinates and to inspire their intense loyalty. Finally, in the words of Douglas MacArthur: "His greatest point was his strength and firmness of character." MacArthur added that Pershing was not a strategist or tactician and that he was not as smart as March.

On May 28, 1917, within a month after the notification of his momentous assignment, Pershing sailed from New York with a small staff. His closest confidant and Chief of Staff, Harbord, a newly promoted lieutenant colonel, was perceptive about the challenge facing him and his fellow officers. He wrote after he had been at sea three days: "Officers whose lives have been spent in trying to avoid spending fifteen cents of government money now confront the necessity of expending fifteen millions of dollars,—and on their intellectual and professional expansion depends their avoidance of the scrapheap." At that time neither Pershing, Harbord, nor anyone else foresaw the great expansion that did take place in the next seventeen months. The Army's strength would reach about 3.7 million, and it would take more than 200,000 to officer this huge force. In turn some two million officers and men would serve under Pershing's command in the AEF before the Armistice in November, 1918. What this meant was that many middle-aged field grade officers suddenly found themselves rapidly going up the promotion ladder. After a career of lockstep promotion by seniority within their branch (prior to 1890 it had been within the regiment), they welcomed the opportunity that selection on the basis of merit brought.

Harbord's career is an example of the good fortune enjoyed by some. Commissioned in 1891 after two years in the ranks, he had served as a major of volunteers in 1898. When the Spanish War came to a close, he reverted back to his first lieutenancy. In 1917, he was fifty-one, a student at the Army War College and a major with date of rank of December 10, 1914. In May, he became a lieutenant colonel; in August, a colonel; and in October he put on a star. After he left Pershing's staff in the spring of 1918, he commanded a brigade, and after he became a major general in July, 1918, a division.

A similar advance took place on John L. Hines' promotion record. A West Pointer in the Class of 1891, Major Hines had accompanied Pershing into Mexico as his adjutant. As in the case of Harbord, he also received promotion to lieutenant colonel in May, 1917 (he outranked Harbord as a major by more than two years). He went with the Pershing party to France as the assistant adjutant general and also received a promotion to colonel in August. That fall, "Birdie" Hines left the staff and began one of the most remarkable careers as a combat commander in the AEF. He led with outstanding success a regiment, brigade, division, and corps in the last year of the war. Commensurate with these commands, he wore a star in April, 1918 and became a major general in August.

Harbord and Hines had exceptional careers in the AEF but theirs were not so dissimilar from other Regulars' promotions. In the late summer and fall of 1918, the Army promoted to brigadier general several officers who had been captains when the war began. Some thought that a disproportionate number were field artillery officers (among them, William Bryden and Robert M. Danford), which provoked some suspicion that the Chief of Staff, a former field artilleryman, was partial to his branch. The Chief of Staff pointed out that his rationale was to give field artillery officers command of field artillery brigades. At that, one officer outstripped the artillerymen in promotion. The commander of the first aero squadron in the Army, Benjamin D. Foulois, was a thirty-seven-year-old captain with less than a year in grade when the war came. Because of his specialty, he found himself a major in June and then enjoyed a spectacular leap to brigadier general in September, 1917. The irony of these promotions was that with the end of the war and the reduction in strength of the Army, most of these men who had advanced rapidly went down almost as far—faster.

In the summer of 1917, while Pershing and his staff were

laying the foundation for the AEF, the War Department organized the bulk of the divisions that would see combat. During the early days of training, Secretary Baker thought that it would be a beneficial experience if he sent the division commanders and their chiefs of staff to France. This would serve two purposes. For them, it would be an introduction to the war and, presumably, they could pick up all sorts of valuable information in their conversations with Allied division commanders and staffs. At the same time, the visits would give Pershing the opportunity to judge their fitness for command. On their return, Baker himself would interview each general individually and form an opinion as to "their vigor and alertness both of mind and body."

Aside from the fact that this would take the commander and the chiefs of staff away from their divisions in the critical early period, the plan did have merit. Many of these generals who made the two-month sightseeing trip did not meet Pershing's standards. His subordinates, as well as the British and the French, commented on the physical debilities and the seeming lack of grasp various of these generals showed on their tours. In a personal and confidential memorandum Pershing frankly spelled out each man's deficiency, as a few examples indicate:

> *General Thomas H. Barry.* General Barry is past 62 and, although comparatively vigorous, is, I fear, too far along in years to undertake to learn the handling of a division or to stand the arduous work very long. He has never had a large tactical command.
> *General Harry F. Hodges.* This officer has not had any practical experience in commanding troops, as all his life has been spent as a constructing engineer. He is 58 years of age and it is too late for him to begin to learn to be a soldier. It will simply be a waste of time, with no result except failure.
> *General Henry A. Greene.* General Greene is in his 62nd year. He never has been an active man and he is very fleshy. I would recommend that he be not sent back as a division commander.

Pershing concluded this memo with the general comment: " . . . many of these officers will do perhaps to take the administrative control of training camps, but some of them are so old and infirm in appearance that they should not be allowed to have a command, where their infirmities will react upon younger and more capable men." In this last category, Pershing placed a former Chief of Staff, America's most famed soldier, Leonard Wood.

Doing so led to a controversy with political overtones, since Wood was a close friend of TR's and had been considered a possible opponent of Wilson in the 1916 Presidential election. March, Baker, and Wilson, however, backed Pershing and were criticized. A postwar cartoon in the *North American Review's War Weekly* called up memories of Wilson's campaign slogan of 1916 when it showed Wood saying to Roosevelt: "Well, he kept *us* out of the war."

During the war—particularly in the winter of 1917–1918 and the following spring—there raged the greatest controversy between the Allies and the Americans. This touched significantly on the quality of the American officer corps. After a series of disasters in late 1917, the grinding to an inconclusive halt of the British offensive on the Western Front, and the German action at Caporetto, and the Russian Revolution which meant the withdrawal of Russia from the war, the Allies began to think seriously of the need for a great reinforcement of manpower by the Americans. Logically, fewer men would be needed and they could all be combat troops if they served as replacements in British and French units. This method would obviate the necessity for garrison commanders and staff officers as well as American logistical troops. It seemed a dangerous gamble to await the transportation of the larger number of officers and men required for an independent army, to say nothing of the lengthy time it would take for commanders and staffs to gain experience. But Pershing stood firmly against amalgamation, again with the support of the War Department and the White House.

Meantime, Pershing went about the business of building an independent force. Throughout 1917, the AEF increased slowly;

it was not until the spring of 1918 that the transports began to speed up. Although Pershing was determined to have an independent army (indeed his orders so stated), he was willing to have the French and British play a large role in training his army. His program, in brief, was to divide training into three phases. He assumed, correctly, that the troops would get little training in the States. (In fact, the War Department took large drafts from the divisions in the training camps in order to bring other divisions up to strength at the ports. This meant a constant, large-scale turnover in the divisions, with the obvious damage to training.) French and British instructors either gave or supervised instruction in the first phase of weapons training and tactical exercises up to division level. Since the automatic weapons and field artillery were of Allied make, the need for their help in weapons training was evident. They also showed how to construct trench systems and demonstrated techniques of trench warfare. In the second phase, the Americans went to the front, where they served as subordinate units within and under French or British command. Frederick Palmer, the noted war correspondent, commented on how carefully the French treated the first American units to reach the front: "We were nursed into the trenches with all the care of father teaching son to swim." The final phase consisted of division maneuvers, concluding with the division going to the front as part of a French corps.

Allied tutelage became burdensome, and so did the close supervision of Pershing's staff. The Chief of Staff of the 2nd Division, who later commanded the 3rd Division in the fall of 1918, Preston Brown, indicated his problems in both regards in this story: "Shortly before the jump off at St. Mihiel, a young lieutenant, 24 or so, came to his headquarters from GHQ. He asked, 'Is there anything I can do for you?' P. Brown answered, 'Since we've been here, we've had to fight the British, French, Italians and Belgians before we could fight the Germans. You can get the hell out of here.'"

Pershing did not want the Allies to have too much control or influence over his troops. He also wanted to keep a firm hand, through his staff, on his army. This is one reason why it is difficult to evaluate the abilities of the commanders who served under him. First under the Allies then under Pershing, they were given little discretion. Hunter Liggett as a field army commander in the last month of the war evidently did gain a good deal of independence from this close supervision, but his army and the various corps staffs were keeping the divisions under a rather tight rein. The type of warfare contributed to this. As the Americans began to get out of the trenches and go on the offensive the tight control continued, but it is doubtful if it would have over a long period of open warfare.

Time, itself, is a key problem in evaluating division commanders. Most had little opportunity to show their abilities because the Americans were not heavily engaged until the last six months of the war. The first regimental assault took place on May 28. And more than 1.3 million of the two million men in the AEF arrived in the five months between that action at Cantigny and the Armistice. The First Army did not go into battle until September 12 at St. Mihiel, two months before the end of the war.

Then, just as there were great turnovers in personnel in the divisions, there was a large turnover in division commanders. Of the twenty-nine divisions that saw combat, three had four different commanders; eleven had three; nine, two; and only six kept the same commander. This does not include *ad interim* commanders. Since twenty-one of these divisions were formed in the summer of 1917 and the other eight in the following fall or winter, it is evident that few generals had very much time in command, let alone leading the division in combat.

Good commanders surfaced as well as did some bad ones. Pershing was ruthless in relieving those who failed. In the midst of the Meuse-Argonne campaign in October, 1918, he replaced a corps commander, three division commanders, and several brigade commanders. He even had the objectivity to see that he was trying to do too much as C-in-C, AEF and Commanding General, First Army. He called on Liggett to take over the second job. This was a wise move. The new commanding general rested the First Army and prepared it for its last battle, which turned into a breakthrough in the first week of November.

In assessing American commanders in the war, one must return to that point about time. The United States was in the war nineteen months and, of that period, the AEF participated on a large scale in combat for less than six months. In that brief period, Liggett, Hines, Charles P. Summerall, and a few others did demonstrate clearly their skill as commanders. Others— such as Harbord, who had a reputation as a successful brigade and division commander—do not appear so favorably after a study of the records. This is not to say Pershing was not a great commander; but Pershing himself did not show any particular skill as a strategist or tactician. For his leadership in developing the AEF from less than a hundred thousand to more than two million he deserves the highest tributes. He was a superb military manager and he was always the commander—every doughboy knew that.

Just as Pershing built the AEF, Peyton C. March invigorated the War Department General Staff and spurred it to a high level of efficiency in the last eight months of the war. Unlike Pershing, March came to power after his organization had weathered severe shocks and great changes. He had to break established patterns and bring powerful and reluctant bureaus into line for the war effort. Lear also knew March in the pre-World War I period, and he thought him as harsh and as abrupt with subordinates as Pershing. He was, if anything, more so. He set the tone of his ruthless but efficient and effective administration as Chief of Staff when he turned on an officer who was presenting papers to him. When this man gave him a document concerning relief of a general, March asked him why he was showing him this. The officer replied it was because he understood that the general in question was a friend of the new Chief of Staff. March exploded: "Don't ever bring in another paper to me because you consider that it concerns some[one] who you think is a personal friend of mine. I want it distinctly understood by you and by everyone else in the office that as Chief of Staff I have no friends." Douglas MacArthur pointed out another of March's attributes: "He was a man of principle. When he said—'No'—he said 'No'—without making black seem whitish or white seem blackish. . . . When you left March's office, you knew exactly what he wanted."

There was a good deal of friction between the two strong men—March and Pershing—whose perspectives were naturally different as Chief of Staff and as the C-in-C, AEF. One of the most irritating problems that came up was the handling of the promotions to brigadier and major general. The first confrontation came within a month after March, who had served in France as Pershing's Chief of Artillery, had returned—a major general—to be Acting Chief of Staff. He asked Pershing for recommendations for the generals' list. When the nominations appeared, only three of the ten officers Pershing had recommended for promotion to major general and less than half of those names he had sent in to be new brigadier generals were on it. To make matters worse, one artillery colonel—a friend of March's then serving in the AEF—was on the list even though Pershing had not recommended him. Pershing, a four-star officer, complained to March—still a major general. March's response was that he had to think of the Army as a whole and that he valued Pershing's recommendations no more and no less than those of other commanders.

Again in the summer and the fall of 1918, this same issue arose. The fact that March was by then no longer merely Acting Chief of Staff and had become a full general did not help matters. Pershing was accustomed to having his recommendations complied with. Beyond that, he found it intolerable when March continued to promote an occasional AEF officer without his recommendation, including Douglas MacArthur in particular.

While March was right in considering the Army as a whole, he should have shown more tact and awareness of Pershing's special position. The issue was never resolved. Promotions stopped on Armistice Day. And in the months following the war, AEF generals were bitter toward March when he promptly reduced them to their regular ranks as their units returned to the United States to be demobilized. At the same time officers in the War Department who still held emergency slots kept their temporary rank.

Both March and Pershing were military managers, but the public showed more interest in the heroic-leader image—to borrow terms from Morris Janowitz. As commander of the AEF, Pershing projected the image the public wanted. Congress responded by honoring him with the rank of General of the Armies. March, in turn, was available as a scapegoat for complaints against the wartime Army (most of which ironically concerned the AEF) which were ventilated after the war. Congress reflected this in its refusal to promote March to permanent four-star or even three-star rank. Thus he reverted to his regular major general's rank in June, 1920, and left office and retired a year later to oblivion as far as the public was concerned.

The role of American command and commanders changed in 1917-1918. The leading generals showed their skills as organizers and managers. There simply was not enough time for commanders to lead their divisions, corps, or armies long enough to evaluate them as long-term combat commanders. Yet it was difficult for both soldiers and civilians to accept that the outstanding American generals were what could be called desk officers. Indicative of this lack of understanding is that one Congressman in the spring of 1918 suggested that staff officers on duty in Washington be required to wear a white band on their sleeves. By the same token Harbord, who commanded the 2nd Division less than two weeks, including a total of two days in combat, was considered by Pershing and others a top-notch division commander—even though what happened to that division at Soissons under Harbord's command indicates that his talents were much better used as the logistical manager of the Services of Supply.

COMMANDERS IN WORLD WAR I

In World War II both soldiers and civilians were more sophisticated, so there was not the same lack of understanding of the role of a military commander. Then the war lasted long enough for a real testing of combat commanders. For these officers, that time of great transition from a Civil War army to a modern army—World War I—was a valuable training ground.

Bibliography

The author has conducted extensive personal interviews. Among the original source collections consulted for this essay were the Baker, Pershing, and Hugh Scott papers at the Library of Congress and transcriptions of interviews and reminiscences of Duncan, MacArthur, Simpson, Lear, Bradley, and Foulois in the author's files.

Coffman, Edward M. "The American Military Generation Gap in World War I: The Leavenworth Clique in the AEF" in William Geffen, ed., *Command and Commanders in Modern Warfare*. Colorado Springs: USAF Academy, 1969.

Coffman, Edward M. *The Hilt of the Sword: The Career of Peyton C. March*. Madison: University of Wisconsin Press, 1966.

Coffman, Edward M. *The War to End All Wars: The American Military Experience in World War I*. New York: Oxford University Press, 1968.

Hagedorn, Hermann. *Leonard Wood: A Biography*. 2 vols., New York: Harper, 1931.

Harbord, James G. *Leaves from a War Diary*. New York: Dodd, Mead, 1925.

Hart, Sir Basil Liddell. *Reputations Ten Years After*. Boston: Little, Brown, 1928.

Finnegan, John P. "Military Preparedness in the Progressive Era, 1911-1917," unpublished Ph.D. dissertation, University of Wisconsin, 1969.

Janowitz, Morris. *The Professional Soldier*. New York: Free Press, 1964.

Kreidberg, Marvin A. and Henry, Merton G. *History of Military Mobilization in the United States Army, 1775-1945*. Washington: Government Printing Office, 1955.

Liggett, Hunter. *AEF: Ten Years Ago in France*. New York: Dodd, Mead, 1928.

Mott, T. Bentley. *Twenty Years as Military Attaché*. New York: Oxford University Press, 1937.

Palmer, Frederick. *America in France*. New York: Dodd, Mead, 1918.

COMMANDERS IN WORLD WAR I

Pershing, John J. *My Experiences in the World War.* 2 vols., New York: Stokes, 1931.

Pogue, Forrest C. *George C. Marshall: Ordeal and Hope, 1939-1942.* New York: Viking, 1966.

Quirk, Robert E. *An Affair of Honor: Woodrow Wilson and the Occupation of Vera Cruz.* Lexington: University of Kentucky Press, 1962.

Scott, Hugh L. *Some Memories of a Soldier.* New York: Century, 1928.

Smythe, Donald. *Guerrilla Warrior: The Early Life of John J. Pershing.* New York: Scribner, 1973.

Stimson, Henry L. and Bundy, McGeorge. *On Active Service in Peace and War.* New York: Harper, 1947.

Weigley, Russell F. *History of the United States Army.* New York: Macmillan, 1967.

IV

The Composition of Armed Forces

Conscription and Voluntarism:
The Canadian Experience
R.H. ROY
page 200

The Multi-cultural
and Multi-national Problems
of Armed Forces
RICHARD A. PRESTON
page 225

The Army of Austria-Hungary,
1868–1918: A Case Study
of a Multi-ethnic Force
GUNTHER E. ROTHENBERG
page 242

COMMENTS

Students of military systems have argued that armed forces so directly mirror the societies they serve that an army can be no healthier than the society that nurtures it. Many of them have also argued that in order to generate optimum strength and effectiveness, armed forces ought to be deliberately designed to make them as complete a reflection of their societies as possible. In their composition, they should include a cross section of the whole population they serve; then they will best be able to establish that rapport with the whole society from which they can draw on the whole society's strength.

This argument was heard especially frequently during the late 1960s, when Americans debated whether to relegate conscription to a stand-by role for extreme emergencies and to substitute in peacetime a volunteer army for a conscript one. Among the many variations on the argument, special prominence was given to the charge that a volunteer army would become a separate military caste, a sort of praetorian guard, widening the gulf between the soldier and the civilian and thus aggravating misunderstandings and tensions between the military and the larger society. This process, opponents of the volunteer army said, would reduce military effectiveness as well as hasten the growth of an already dangerous American militarism.

There is an undeniable historical correlation between conscript armies and political democracy. The idea of the **levée en masse** was a product of the democratic revolutions of the eighteenth century. A people's army drawing on the whole potential military manpower of the nation as citizen-soldiers can offer a mighty defense for popular government against external foes—as the armies of revolutionary France demonstrated at the outset. In addition, a people's army can scarcely threaten the people's own government; in a more recent French example, it was the unwillingness of French conscripts to obey orders of their professional officers to strike against the Fifth Republic *that was crucial in saving President Charles de Gaulle from having to face an attempted military coup in metropolitan France when he abandoned the struggle to retain Algeria.*

In the United States, the theory of the military obligation of citizenship has been a principle of popular government from the beginning—from the colonial militia systems even before there was a United States, and from the reiteration of the colonial tradition of a universal obligation to military service in the new state constitutions of the Revolutionary era, the recognition of the militia tradition in the United States Constitution of 1787 and the Bill of Rights, and the re-establishment of the universal-service doctrine in Federal law as early as the Militia Act of 1792. Throughout American history, the citizen army remained associated with the republican, Jeffersonian political tradition, and an emphasis on a professional army with the Federalist, Hamiltonian political tradition.

But too much of the debate over a volunteer army was conducted on a level of principle and ideology. While the opponents of the volunteer army emphasized the historical connection between the citizen-soldier and democracy in theory, they paid much less attention to the mundane fact of the matter, that throughout most of American history, a universal obligation to military service was only a theory, and the United States secured its military defense from a volunteer army—without any particular danger to democracy. In practice, the United States had had a professional army for a long time, without noticeable adverse effects on democracy. To be sure, the Army had been small, and even a volunteer army in the late twentieth century would have to be large; but granting that change in circumstances and the importance of the change, there was still reason for a more matter-of-fact examination of the realities of experience with a volunteer army than usually occurred during the debate.

COMMENTS

R.H. Roy's discussion of the Canadian experience with volunteer and conscript forces, a lecture given in the autumn of 1973, is distinguished by its matter-of-factness. It deals less with principles and ideologies than with political maneuvering and specific events. The Canadian situation does not closely parallel the American one in matters of a volunteer army, because the heart of the Canadian problem with conscription has been the presence of the great French linguistic and cultural minority. But the evident vitality of Canadian democracy despite the virtual impossibility of Canada's creating a nation in arms is worthy of notice, when we have often heard that democracy demands as its military system the nation in arms. R.H. Roy is Professor of Military History and Strategic Studies at the University of Victoria, British Columbia.

Conscription & Voluntarism: The Canadian Experience

R. H. ROY

CANADIANS have sometimes been called an "unmilitary people," usually by military historians within Canada's national boundaries rather than without. In large measure this is true—except in wartime. In the two major wars in which we have been involved, the Great War and the Second World War, Canada's armed forces have made a very creditable name for themselves. In 1914-1918, the Canadian Corps, fighting as part of the British Army in France and Belgium, made a reputation for itself on the Somme and at Vimy Ridge which gained its recognition as one of the elite corps among the Allied forces. Our Navy was almost nonexistent at that time, and although there was no Canadian air force, approximately one quarter of the fighter and bomber pilots in the Royal Flying Corps and the Royal Air Force were Canadians, who, incidentally, ranked among the top aces in the air battles of that war.

In the Second World War Canada had a little over one million men and women in the three services. For a country with a total population at that time of some eleven and a half million people, this was a creditable showing. In this second war Canada had a navy of some 500 ships ranging from motor torpedo boats to aircraft carriers, an army of over half a million men, and an air force which, once again, was merged into the Royal Air Force and whose squadrons fought from Burma to the Western Desert and from Britain to the Aleutians.

THE COMPOSITION OF ARMED FORCES

During these two wars by far the greater proportion of those who served overseas—95 percent or more—were volunteers. But conscription is by no means unknown in Canada, and indeed the very mention of the word was enough to make some politicians shake, and the implementation of it could bring governments to the brink of resignation. Conscription was to bring two crises in Canada's political history that even today can work up emotions and lead to heated arguments. Thirty years ago, in some areas of Canada, the issue might also lead to blows, and fifty years ago, in one or two minor incidents, it did result in bloodshed.

To understand Canadians' attitudes towards voluntarism and conscription in this century, one cannot overlook the historical background of the nation. The 150-year tenure of France's possessions in Canada ended with the capture of Québec in 1759 by British forces under Major General James Wolfe. The victory on the Plains of Abraham brought under British control a population of about 65,000 French-speaking citizens, almost all of the Roman Catholic faith. A few hundred took the opportunity offered them to return to France, primarily those who held official positions in the civil service. Aside from those few, the remainder, almost all of whom followed rural pursuits or were engaged as small businessmen in the fur, fish, and forest trades, stayed in Canada. Like their English counterparts to the south, time and distance had played their part in creating a different outlook on life from that of the mother country across the Atlantic.

Three events in the next half century were to play major roles in shaping not only Canada's future, but particularly Canada's military role in the years to come. Sixteen years after Québec's capture—a short period in our colonial history—the English colonies south of Québec broke away from Great Britain, and for the next seven years it was sometimes touch-and-go whether the Revolution would succeed or not. The unrest in the American colonies prior to 1776 had not gone unnoticed by the military commanders in British North America, especially in

Québec. The anticipated migration of English colonists to the north had never materialized after 1760, and as a result the governors in Canada were more than lenient to the new subjects of His Majesty. Basically the French retained their language, their religion, their culture, and in a word, their society. In return it was expected by the British that the clergy and the seigneurs would remain neutral should a revolution to the south occur, and possibly even that they might take up arms to defend Québec should an invasion be launched against it. The attitude taken by the British paid dividends. The French Canadians remained either neutral or passively defensive. Rarely if ever did the British commanders operating from Québec have to look over their shoulders to worry about French Canadians. Indeed, the demands of the British forces gave a considerable stimulus to the growing settlements along the banks of the St. Lawrence.

The American Revolution was followed a decade later by the French Revolution, and here again events were in train that affected Canada. British North America not only survived the American Revolution but gained immensely from it. For the first time thousands of English-speaking colonists came to the Maritimes and southern Québec, the latter influx being so considerable that a new province was created—later called Ontario—to enable the United Empire Loyalists, as they were called, to enjoy the political, legal, and other institutions to which they were accustomed. But while "English Canada" was growing, so too was the French Canadian population. However, the overthrow of the French monarchy, the execution of the aristocracy, and more than anything else, the overthrow of the power of the Roman Catholic Church by the French revolutionary government had a major impact on the people of Québec. There was no sympathy among French Canadians for the French Revolution and its principles. Indeed, the French Canadian clergy looked with horror upon the events in France, and their influence, together with their control over the education in Québec, weakened still further the ties between Old and New France.

THE COMPOSITION OF ARMED FORCES

A third critical event came towards the end of the Napoleonic era. There is no need to go into the causes of the War of 1812 between Great Britain and the United States, but its impact on Canada does deserve more than passing mention. When the war broke out, there were about 95,000 English-speaking colonists in Ontario and some 335,000 colonists in Québec, the greater proportion of whom were of French descent. Ontario was to be the scene of most of the battles, and English Canadian militiamen fought loyally and well beside the few British regiments of regular troops in the various battles and engagements in which they were involved. Although they were further from the area of conflict and consequently not as involved in active campaigning, French Canadian militia units were involved in the war. Their valor in the field and their service in garrison and other duties while serving under King George equalled that of their grandfathers who had fought under Louis XV of France. For the first time French- and English-speaking militia units fought side by side with British regulars to defend their homeland against a series of invasions from the south over a considerable length of time. Moreover they were successful. Their success, to quote Professor A.R.M. Lower, laid the foundation of Canadian nationhood, for not only did it exemplify the determination of the two founding nationalities to work and fight together side by side, but it seemed to prove that Canada could be preserved by a united effort even against the powerful republic south of the border. It also left a suspicion that the only military threat to the developing nation would come from the south. The Fenian raids of 1866, together with periodic outbursts by enthusiastic American politicans about the "manifest destiny" of the United States, periodically reinforced that suspicion, which was to linger on until the turn of the century.

Canada achieved nationhood in 1867. One of the factors leading to it was the Fenian threat. The raids by these Irish Americans over the Canadian border, fortunately, struck at both Ontario and Québec, which gave weight to the argument by the confederationists that only in unity could the strength of the various colonies be effectively utilized. And unity in Canada in the 1860s, especially political unity in the central provinces, was becoming so fragile that political deadlock appeared imminent.

What had happened? In the half century since the end of the War of 1812, British North America threw open her gates to immigrants from Europe, and decade by decade Canada's population increased steadily. But with each increase, the proportion of French- to English-speaking Canadians grew steadily less. By the 1850s Ontario's population for the first time became larger than Québec's and by 1867, the addition of Nova Scotia and New Brunswick added to the English-speaking element. A few years later, when Manitoba, British Columbia, the North-West Territories, Prince Edward Island, and ultimately the new prairie provinces became part of the Dominion, the national origins of the total population definitely established the earlier trend of domination by those of British extraction.

There is little doubt that, in the nineteenth century and even in the twentieth, there were many English-speaking Canadians who would not have grieved if their French-speaking compatriots had declined in numbers and in influence. This was the fear of many French Canadians as well, and they were determined to maintain their culture, their religion, their language, and their way of life even though they lived on a continent that was predominantly English-speaking. Large families in Québec increased the number claiming French origin from 65,000 at the time of the conquest to over a million at the time of Confederation. This number had doubled by the time the Great War broke out, and by 1939 approximately three and a half out of a population of eleven and a half million people in Canada were either French-speaking or of French descent. Comparisons are odious, John Donne said, but in historical perspective an American might reflect back to 1941 and consider what the impact might have been on the United States if, in a huge triangle running from Florida to the Dakotas to Southern California, there was a population of 40,000,000 French-speaking Americans, dominated by their clergy, strong in their traditional culture, insular

THE COMPOSITION OF ARMED FORCES

though not isolated, and prone to vote as a bloc in times when they feared their way of life threatened. In roughly proportionate terms, that is what is termed in Canada the "French Canadian fact."

How and in what ways did the two founding national groups complement and compare with each other in matters of defense? Historically each group had been accustomed from the early colonial period to being subjected to military service. In New France the French Canadian *habitant* owed military service to the governor of the colony, and through him to the monarch. Periodically in times of danger he was called upon to bear arms against the Indians, the English colonists, or the regular forces of the British crown. After the conquest, aside from a joint colonial effort against Pontiac, no effort was made to compel the French to fight against the revolting English colonies, but the concept of the *"levée en masse"* remained the law of the land. In brief, during the British colonial period, both French- and English-speaking colonists were subject to the militia laws, which gave the government power to call upon all healthy males between sixteen and sixty years of age to aid in the defense of the country in time of war and engage in military training in time of peace. Since Canada was a colony and, consequently, was garrisoned by British troops, it was assumed that the militia would fight in support of the British regulars. It was also quite natural that the Canadian militia would mirror the British in their uniforms, drill, training, structure, and so forth. The British regimental system was adopted and remains with us today. The militia trained with British-supplied arms, copied its tactics from British army manuals, adopted the British rank structure, saluted the British flag, swore an oath of allegiance to the British sovereign, assumed British-style names for its militia battalions and, in a word, absorbed as standard practices those with which it was both familiar and comfortable. As Canada continued to expand with the completion of the transcontinental railway in 1885, and as immigrants from Britain and Ontario began to fill in the prairie and Pacific provinces, it was only natural that newly created militia units should follow the example of those to which the young men were familiar in the East. This would have occurred in all probability even had the recently created Ministry of Militia and Defense not imposed the system on them for the sake of national unity and military efficiency.

In Québec, and one must remember that not every resident of *"la belle province"* is French-speaking, the militia remained on an adequate though not necessarily flourishing basis of the post-Confederation era. The first Minister of Militia was a French Canadian, and in the next thirty years, one of the outstanding ministers, who was to occupy that position for well over a decade, was another French Canadian, Sir J.P.R. Adolphe Caron.

From the outset service in the militia was on a voluntary basis, even though the right to impose a *"levée en masse"* remained on the statute books. The reasons for the voluntary method were not difficult to find. After the scare created by the Fenian raids, relations between the United States and Canada improved steadily, especially after the Washington Treaty of 1871 which resolved many of the outstanding differences between the United States and Great Britain. Without the stimulus of a threat, there was no need to maintain a large force, even though the British withdrew their garrisons in 1870-1871. Another major reason was the expense that the imposition of any form of conscription would involve. Many members of Parliament complained about the costs involved in maintaining the tiny permanent force of regulars which was created to take the place of the withdrawn British forces. At that, the Canadians had no need to establish bases from scratch, as they took over and occupied former British barracks, gun positions, and other military establishments that were given to the new nation.

The call to keep Canada's military establishment at the lowest possible level owing to the financial stringency of the federal government was to be heard time and time again, and as this consideration played a major part in establishing and maintaining the voluntary system in Canada, it deserves some attention. It

THE COMPOSITION OF ARMED FORCES

was almost twenty years after Confederation before Canada built a railway connecting the East with the West. The need to unify the country geographically and economically presented a tremendous challenge to the new nation and, at the same time, saddled it with a seemingly insurmountable, ever-increasing debt. When in the 1870s and 1880s Great Britain suggested that Canada might assume more military responsibility for her own defense, Canada invariably pointed to the millions she was paying toward railway construction which, when completed, would help bind the Empire and serve a very useful strategic purpose in imperial defense. Railways were but one expense. The country was immense, and many parts of it had not even been surveyed. There was a cry for roads, canals, bridges, port facilities, policing of the far North, and a thousand other basic needs to create an infrastructure for national growth.

Canada's militia, therefore, suffered as a result of national growth in the latter decades of the nineteenth century. As the railroad pushed into what are now the prairie provinces at a time when the buffalo were nearing extinction and a parsimonious government felt able to ignore the desperate plight of the natives, the North-West Rebellion broke out, when Indians and half-breeds took up arms under the leadership of a French Canadian *métis*, Louis Riel. It was the most serious situation facing Canada between the War of 1812 and the outbreak of the Great War and resulted, ultimately, in 7,000 militiamen being sent out to the West to quell it. All of these men were volunteers, who came from almost every province in the dominion. So unprepared were they for serious action in the field, however, that Sir Adolphe Caron had to rely on two major private national organizations to maintain his forces during the campaign. During the summer of 1885, the Canadian Pacific Railway and the Hudson's Bay Company had the responsibility of transporting and supplying the militia, providing communication between Ottawa and the various headquarters in the field, arranging purchases for the force from local sources, and in general taking care of all the logistical needs of the North-West Field Force. The Rebellion did result in a mild improvement in the militia, but the hanging of Louis Riel, the leader of the rebellion, was to arouse a storm of protest in Québec which, in the long run, was to more than offset the slight gains made in the overall strength and efficiency of the militia.

In 1914 Canada had a regular army of about 3,000 men and a militia or reserve force of 55,000. The latter figure had increased by about 20,000 from the turn of the century. It was a reflection partly of the increase of Canada's population from under five million in 1891 to well over seven million by war's outbreak, but more especially it reflected the enthusiasm of two Ministers of Militia together with a series of reforms and improvements made in the militia itself. Coupled with that was the impact in Canada of what might be termed the imperialist idea or, perhaps better, the vicarious show of pride among Canadians who shared the pleasure in the accomplishments and conquests of the British Empire at the turn of the century. There is no way of measuring this feeling, and very little if anything has been written on it, but an examination of the origins and growth of many Canadian militia regiments in the two or three decades before 1914 leads me, at least, to believe that this was one of the factors that must be taken into account to understand why the voluntary method worked as well as it did.

It might be well at this point to slip a Canadian militia unit under a "microscope" and examine it briefly. Each has its own characteristics, but there is more similarity than difference among them. The moving spirits behind the origin of a militia unit were usually a very small group of "prominent citizens" who, for various reasons, felt their community should have such a unit and believed the local population would support it. The conditions existing, a public meeting would be called, the plan outlined, and the young men present requested to sign a petition requesting the District Officer Commanding to forward to Ottawa his support for the establishment of a unit. If he agreed, and if the military authorities in Ottawa agreed to the proposal, a Service Roll would be sent to the chairman of the meeting for

THE COMPOSITION OF ARMED FORCES

signatures of those willing to enlist for service. At the next meeting this would be done, and at this time also nominations were put forward for those who should fill the commissioned and even senior noncommissioned ranks. Frequently, though not always, these would include those who had originally proposed the idea, but Ottawa would always have the final word, usually on the advice of the district headquarters.

Such were the origins of the British Columbia Dragoons, for example, whose recruiting area was the Okanagan Valley. Several decades passed, however, between the first petition and the issuance of authority to raise the first squadron of cavalry. Those wanting the unit claimed that it was needed originally because there were more Indians than white men in the interior of the province, and later, that such a corps would prove useful should the need arise to maintain law and order among a recent large influx of miners, "generally a rough class liable to strike at any time." Later appeals to Ottawa pointed out the value of a militia force in the interior "in case of any trouble with our neighbours south of the border line" which might result in the cutting of the transcontinental railway line. Neither petitions nor political pull were to have any effect on loosening the financial restrictions imposed on the Department of Militia and Defense. It was not until 1908 that the first squadrons were organized, by which time the wartime experiences of men who fought in the Riel Rebellion, the South African War, and in various brushfire wars throughout the empire were readily available among those who joined the regiment.

The reasons why the volunteers enlisted in this period are probably as numerous as the volunteers themselves. There seems to be little doubt that the militia units were popular in the community, and part of that popularity was based on the similarity of the Canadian units with their British counterparts. Tens of thousands of British immigrants who came to Canada in the nineteenth and early twentieth centuries had served with the British Army, and many of those who were active in organizing, promoting, or supporting the Canadian units were former British officers. The literature of the day, ranging from the *Boys' Own Annual* to the works of such authors as G.A. Henty and Rudyard Kipling, helped to prepare young men for service to God, King, and Country. The tales of adventure with their background laid on the North-West Frontier of India or in deepest Africa captured the imagination of thousands of young men and were read as keenly in Canada as they were in Britain. Both in schools and in churches, the theme of duty was hammered home and came to be accepted without question. That the soldiers of the Queen were performing both a civilizing and a Christian mission in expanding British control over the benighted savages and less-enlightened peoples in far-distant lands were patently obvious. If God wasn't an Englishman, at least he spoke with an English accent.

The appeal of the scarlet tunic, then, to the young men in the Okanagan and elsewhere was considerable. There was a vicarious pride in being even somewhat remotely associated with those who were bringing the benefits of civilization to those "beyond the pale" or guarding the distant frontiers against the encroachments of rivals. Hand in hand with this, of course, was the prestige and the measure of social distinction which went with participation in a uniformed force, especially if the unit was well led and well trained and had a reputation for being "smart." If it had a good military band which could double as a dance band so much the better, and if additionally it was cavalry and could prance by the reviewing stand or sweep over neighboring meadows in mock battles, better still. The sight of one's boyfriend, immaculate in his uniform, jingling by in formation with his lance, sword, and carbine, was enough to make any heart flutter, and an invitation to a regimental ball or a regimental sport-meet was sought after as much as an invitation to Government House or the mayor's reception. In that era, before the radio and television provided entertainment in the home, and when the motor car was still an innovation, local militia units provided a social and sports club in many towns and cities. The armories, whether built or rented, were often one of the first community centers, and regimental dances, which incidentally would not be held in church halls, offered plenty of color and excitement to the youth of that time. Regimental picnics, regimental turkey-shoots, regimental maneuvers, the annual summer camp, shooting on the indoor and outdoor ranges with government-supplied ammunition, building a regimental skating rink in winter—in general what we would term today the "community involvement" of the Canadian militia units seems to have had great appeal and accounts in large measure for the continual willingness of the volunteer to offer his services and become one of what a sociologist would term an "in-group."

All of this, however, did not come easily. Although government expenditure on the militia increased greatly in the pre-war years, no one can claim that the militiaman joined up for the pay he received. It was not at all unusual for a militiaman to return to regimental funds the pay he received during the year, except for the period when he attended summer camp and could expect no remuneration from his employer. The same held true for the officers; indeed when commissioned in the militia any young man expected to pay a fair amount for his uniforms, his mount, his mess dues, and occasional contributions to raise money for regimental needs. Ottawa might authorize a unit to have a band, but there might be a waiting period of several years before the band instruments were provided. Similarly a regiment might be authorized to increase its strength by adding to it several companies or squadrons raised in nearby towns, but it was up to the unit to find the money to rent a substitute for a drill hall or armory. Even the amount granted the unit to pay for ten or fourteen days training during the year would never in fact pay for the actual time put in voluntarily by all ranks to bring themselves up to a higher state of efficiency. Militia Headquarters, for example, expected the troopers in the B.C. Dragoons to provide their own mounts and paid for them only during summer camp. Any militiamen, in brief, who expected that an understanding government would show its appreciation financially for their services were quickly disillusioned. The measure of their patriotism, their sense of duty, and the interest the various units were able to generate and maintain among the men becomes all the more remarkable when one considers the rate of the growth of the reserve forces in the immediate prewar period. The growing threat in Europe accounts for part of it, but one cannot ignore the leadership and the community support available to the militia both then and in later years as well.

It is almost trite to say that a regiment would flourish under a good commanding officer and soon show the results of a bad one. For a militia regiment, however, good leadership was essential to survival. A regiment might survive one four- or five-year tour of duty by a lazy or incompetent commanding officer. To survive two such tours would call for exceptional qualities on the part of other senior officers in the unit to supply the drive and imagination that were essential to the health of the regiment. A good C.O. not only breathed life into his command, but he sought to bring it financial and political support as well. To ask a local, well-to-do citizen to serve as an honorary lieutenant colonel was to bestow an honor with the understanding that a monetary contribution to the regimental funds would be forthcoming. A local member of Parliament or member of the legislative assembly was also to be wooed for his political support, a support that in turn might swing several hundred votes his way should he be able to prove that his assistance had helped to benefit the unit. The volunteers certainly needed all the outside assistance they could muster, and the dual role played by the commanding officer made his selection of prime importance to the functioning of the entire system.

War imposes a tremendous strain on a nation and tests every fiber of the state's strength. The longer and more intense the struggle, the more likely is the fabric of society to be warped. Few if any nations enter such a struggle with a full appreciation of its consequences, and certainly this was true of the Great Powers, their empires, and their allies in 1914.

Canada greeted the outbreak of war with a flush of patriotism and enthusiasm. The bands, the flag-waving, cheering crowds, and the rush of young men to the colors recalled what was taking place in the other dominions across the seas and in Great Britain itself. Even in Québec, where for some years there had been a growing opposition to Canadian participation in Britain's imperial ventures overseas, the Roman Catholic clergy expressed their approval of the cause and indicated that Britain could expect support in men and money from *"la belle province."* The recognized political spokesman of French Canada and a former Canadian Prime Minister, Sir Wilfred Laurier, came out strongly for all Canadians to stand behind the mother country—an attitude he was to maintain throughout the war. His opponent, the nationalist Henri Bourassa who was editor of *"Le Devoir,"* one of Québec's most respected newspapers, initially gave his support, writing that Canada had a national duty to contribute according to her resources; but Bourassa suggested that she should first take into account what she could actually contribute. With more men offering to volunteer than there were arms or uniforms to equip them, such words of caution were ignored, especially as most people felt the war would be over within a matter of months. It is significant to note, however, that of the 36,000 Canadians sent overseas in October, 1914, the largest proportion was of British birth, the second largest had Canadian origins, and only a small fraction, 1,245, were French Canadians.

Since the conscription crises of both 1917 and 1944 were going to revolve around Québec's attitude to the wars (although by no means in an exclusive sense), a closer if brief examination of French Canada's attitude toward military affairs is necessary. From the beginning of Canada's military forces, some care had been taken to include French-speaking units in both the regular and militia forces. As we have seen, some of the early Ministers of Militia came from Québec. Two of the battalions sent out to quell the North-West Rebellion in 1885 were French Canadian, but once in the field their endeavors were directed against the Indians rather than the French-speaking, Roman Catholic rebel half-breeds. A small contingent served in the South African War. Men in the French Canadian militia regiments had not the sentimental attachment to British traditions that was so apparent among other units, but there were other appeals to their volunteering. Prestige, the prominence in the community of wearing the Queen's uniform, the small amount of money to be earned at camp in depression periods, the involvement in the regiment's social activities, and the chance by senior officers to distribute small favors or attract some political notice—these and similar reasons attracted many. Proportionately, however, Québec's militia strength and influence were relatively low. At the Royal Military College, for example, the instruction given to Canada's future officers was in English. Partly as a result, of the first 1,000 cadets to graduate only 39 were French Canadians. Officers seeking training to qualify them for higher ranks had to attend British military schools—another barrier to French Canadian influence. On a lower level, words of command and instruction manuals were in English. "In 1870 there had been 15 French Canadian infantry battalions and 64 comparable English-speaking units. In 1914 there were now 85 English-speaking battalions and still only 15 French."

The small percentage of French Canadians volunteering for overseas service in 1914 might have served as a warning to the Minister of Militia, Colonel Sam Hughes, but he chose to ignore it. Indeed, this gentleman, with his roots deep in Ontario Organism with its traditional anti-French-Roman Catholic outlook, had managed to antagonize the militia of Québec since assuming office in 1911. It was owing only to intense lobbying in Ottawa by French Canadian officers during Hughes' absence overseas that permission was granted to raise a French-speaking battalion for overseas service in 1915, the famous Royal 22e Regiment. The tale of blunders, stupidity, and missed opportunities by Hughes during the early years of the war and their impact on voluntarism in Québec borders on the tragic. Add to that Québec politicans who began to question the need for Canada to put forth such an effort and who demanded more privileges for French Canadians as the price of their support of the war, and one has the making of a political-military crisis.

As the war proceeded and casualties mounted, the demand

THE COMPOSITION OF ARMED FORCES

for men to fill the constantly depleting ranks of the Canadian Corps increased. Recruiting leagues and committees were formed, patriotic organizations urged young men to join up, and from the press and pulpit the need was pressed home. An historian writing of the attitude of one fairly large denomination noted: "By 1918 the dominant Methodist attitude of the war was a single belief in the literal truth of the words of the Church's wartime hymn: 'The Son of God goes forth to war, A Kingly crown to gain; His Blood-red banners stream afire; Who follows in his Train?'"

Although volunteers came forward by the tens of thousands, by 1917 voices began to be raised urging national conscription. The need for manpower after the battles of the Somme, Vimy Ridge, and elsewhere, the disintegration of the Russian front, the time it would take before American troops could reach the front, and the obvious need to keep the Canadian divisions at full fighting strength led the Canadian Prime Minister, Robert Laird Borden, to announce in May, 1917 that he felt conscription to be necessary and the decision of his Conservative Party to work for its imposition. He would have preferred to form a national government, sharing power and position with the Liberals as a sign of the united determination of both parties to prosecute the war to the fullest extent. Laurier, however, refused to accept the offer. Political leadership in Québec between Laurier and Bourassa was at stake, and if Bourassa could convince Québec that his voice alone should be heard, then French Canada, Laurier feared, would be even more isolated and inward-looking than ever. As a result Borden held an election to confirm the passage of the Military Services Act in August, 1917. Without Laurier, he had persuaded many Liberals to join his case in a Union government; and in the December, 1917 elections, the Unionist government received an overwhelming vote of confidence except in Québec. There French Canadians voted Liberal, but the Liberal Party as a national party representing all segments of the Dominion seemed to be shattered. If French Canadians voted with their hearts, so also did English Canadians.

THE CANADIAN EXPERIENCE

The occasional riots in Québec, the protests from farmers, and the suspicions of organized labor that came as a result of conscription were not of major significance in the long run. The war was drawing to a close by the time the first conscripts arrived in France, and by no means were all the conscripts of French Canadian origin. Their impact on the outcome of the war was minimal. Of far greater importance was the localization of the Liberal Party to Québec and the need for its new postwar leader, Mr. William Lyon Mackenzie King, to rebuild it.

King was to be the Prime Minister of Canada from 1921 to 1930 and from 1935 to 1948. The shattering blow given the Liberal Party by the conscription issue seared itself on his mind, and as he began the long, painful task of reconstructing the party and making it once again representative of all of Canada, he was determined that never again would conscription or anything like it divide the nation on racial lines. Canada's military and external policy during the next quarter of a century can be understood only in this context. In the 1920s and 1930s Canada's armed forces remained extremely small, and one need hardly add, voluntary. In the 1920s there appeared to be no enemy to threaten Canada or, indeed, the Empire. When a threat began to loom in the 1930s, the demands of the Depression were such that Canada's forces remained unprepared and, in numbers, somewhat smaller than they had been in 1914. With consummate skill, King steered the nation away from any commitment that would involve Canada in imperial or European military affairs. Even the League of Nations with its periodic interest in sanctions and intervention could arouse his fears. Isolationism bordering on neutralism, friendly relations with the United States, and the idealistic notion that no statesmen would be so mad as to launch another world war gave him comfort that he and the Liberal Party could continue the task of building a greater and unified Canada.

The coming of the Second World War brought with it all the nightmares that King dreaded. Even before the war started, both political parties declared themselves to be against the imposition of conscription, although the leader of Canada's small Socialist Party was wise enough to add to his agreement with them that "if

THE COMPOSITION OF ARMED FORCES

we really become involved in war, conscription would be bound to come." Once again, however, the feeling was general that large expeditionary forces sent overseas were a thing of the past. By promising that none other than volunteers would be sent overseas, King ensured his support from Québec, and as double insurance, his Cabinet ministers from that province publicly declared that they would resign if conscription were imposed.

The course of the war, unfortunately, was not determined by the Liberals. German successes in 1940 shattered the complacency engendered by the "phony war," and their subsequent victories in Russia and North Africa, coupled with those of the Japanese in late 1941 and 1942, brought home the realization of the need for a total war effort. These reverses brought about a definite change in the public's attitude. On all sides the cry was for equality of service, and slowly King was forced to give way. The National Resources Mobilization Act in 1940 allowed the government to conscript men for service in Canada, and after Pearl Harbor, three complete infantry divisions were raised for home defense. Early in 1942 King held a plebiscite releasing the Liberals from their earlier pledge not to introduce conscription. The vote was 64 percent for giving the government freedom. Significantly, Ontario voted 84 percent in favor, while Québec gave only 28 percent. Release did not mean automatic conscription, but conscription if necessary. It was not until the latter part of 1944 that King was forced to give way and permit the Army to draw upon the large reserves of trained conscripted infantrymen in Canada to reinforce Canadians fighting in Italy and France. There were some minor disturbances in two or three of the military bases in British Columbia, but nothing like the insurrection King had feared. The long if slow movement towards overseas conscription had helped to prepare the way; the war was obviously going to end soon; and the cause of the Allies was undoubtedly a just one. One might add that, although there was a high proportion of French-speaking Canadians in this group, and especially in the hard-core resisters, Québec as a whole had a far better recruiting background in the Second World War than in the First.

220

THE CANADIAN EXPERIENCE

In the era after 1945, Canada has increased her armed forces in proportion to her population and has maintained them, as well as her reserves, on a voluntary basis. The war in Korea together with her decision to join NATO brought Canada's forces up to the 120,000 mark, although recently this figure has purposely been cut back. The appeal to young Canadians has continued to be fairly traditional, but added to that has been a good pay scale, excellent housing and recreational facilities on bases both in the South and in the far North, and—unique for Canadian servicemen—an opportunity to serve abroad in NATO and UN postings. The relationship between recruiting and retention and the economic health of the state continues to interact now as it has in earlier years, but this apparently is being modified by more generous social services on the provincial and federal levels. Each branch of the service continues to seek its own specialists and age groups despite unification, and indeed the impact of unification and integration of the armed services is still a matter of considerable debate. The shape of the potential enemy, however, leads one to believe that, should it be deemed necessary, conscription would be accepted somewhat more readily than in previous wars. Let us hope the question will remain a theoretical one.

Bibliography

Although a considerable amount has been written about conscription in the Canadian Army, very little has been written about volunteer enlistment. The best single source for the conscription crisis of 1944 is Colonel C.P. Stacey's *Arms, Men and Governments: War Policies of Canada, 1939–1945* (Ottawa: Queen's Printer, 1970). The only comparable work for the conscription crisis of 1917 is the one-volume official history by Colonel G.W.L. Nicholson entitled *The Canadian Corps, 1914–1919* (Ottawa: Queen's Printer, 1962).

Useful references of the two conscription crises included such works as John Swettenham, *McNaughton,* Vol. III, *1944–1946* (Toronto: Ryerson Press, 1969); Lieutenant General E.L.M. Burns, *Manpower in the Canadian Army* (Toronto: Clarke, Unwin, 1956); J.W. Pickersgill and D.F. Forster, *The Mackenzie King Record,* Vol. II, *1944–1945* (Toronto: University of Toronto Press, 1968); Roger Graham, *Arthur Meighen,* Vol. I (Toronto: Clarke, Unwin, 1960); Mason Wade, *The French Canadians, 1760–1945* (Toronto: Macmillan, 1955); and J.L. Granatstein, *Conscription in the Second World War, 1939–1945* (Toronto: Ryerson Press, 1969). A selection of articles from the *Canadian Historical Review* has been collected and published by the University of Toronto Press entitled *Conscription 1917.*

Other accounts are to be found in the biographies of politicians who are involved, such as O.D. Skelton's *Life and Letters of Sir Wilfred Laurier,* Vol. II (Toronto: McClelland and Stewart, 1965); John Dafoe, *Laurier, A Study in Canadian Politics* (Toronto: McClelland and Stewart, 1963); and R. MacGregor Dawson's *William Lyon Mackenzie King: A Political Biography, 1874–1923* (Toronto: University of Toronto Press, 1958). *The Canadian Annual Review of Public Affairs for 1917* (Toronto: Canadian Review Company, 1918) gives a detailed account of the conscription crisis for that year. Unfortunately no similar account is available for 1944.

THE CANADIAN EXPERIENCE

A good account of the background of the anti-conscription feeling in Québec is to be found in Desmond Morton's "French Canada and the War, 1909–1917: The Military Background of the Conscription Crisis of 1917" in *War and Society in North America,* edited by J.L. Granatstein and R.D. Cuff (Toronto: Nelson, 1971).

For the prewar and postwar years, the range and variety of opinions of both the conscriptionists and anti-conscriptionists may be found in the Debates of the House of Commons.

Possibly the best source of material describing how and why Canadians voluntarily joined the militia is to be found in the fairly large number of Canadian regimental and corps histories. The Canadian Permanent Force up until about 1949–1950 was so small that no study has been made of why soldiers joined voluntarily. The list of regimental and corps histories is so lengthy that one can only refer the researcher to the Bibliography of Canadian Regimental Histories published in Charles E. Dornbusch, *The Canadian Army, 1855–1958, Regimental Histories* . . . (Cornwallville, N.Y.: Hope Farm Press, 1959).

COMMENTS

The debate over conscription versus a volunteer army in the United States has featured, almost as conspicuously as the question whether a professional army is inherently subversive of democracy, the charge that an American volunteer army would tend to become disproportionately a black army. Opponents of a volunteer army who make this charge have assumed either that a disproportionately black army would represent still further racist exploitation of blacks, or that a disproportionately black army would be a militarily ineffective army, or both.

How disproportionately black the American armed forces can become is, after all, highly limited by the relative smallness of the proportion of blacks able to meet military enlistment standards out of the total pool of all potential enlistees; the statistics of these populations would restrict blacks to about one-quarter of the armed forces as long as total strength approximated current projections even if all qualified blacks of military age decided to enlist. Still, within such limits, we cannot know how black the armed forces would have become under the volunteer army experiment, had not an economic recession intervened promptly after the inauguration of the experiment, to stimulate white as well as black enlistments. But neither do we know much about what effects a disproportionately high number of blacks in the armed forces would really yield; the public debate and discussion of the issue has dealt more in assumptions and implications than in careful reasoning.

In the autumn of 1971, Richard A. Preston sought to do careful reasoning a service by offering an introduction to the historical background of armed forces that employed troops of differing racial, ethnic, and cultural origins. As he pointed out, the literature on this larger background to current American concern about race in the armed forces is also scant. But Professor Preston contributes a start, based on his long study of British and Commonwealth military institutions, which of course devel-

oped much experience in welding together soldiers of many races. Professor Preston teaches military and British Empire history at Duke University and immediately before going there taught at the Royal Military College of Canada. He is the author of Canada and "Imperial Defense": A Study of the Origins of the British Commonwealth's Defense Organization, 1867–1919 *(Durham: Duke University Press, 1967), and coauthor, with Sydney F. Wise, of the much-used* Men in Arms: A History of Warfare and Its Interrelationships with Western Society *(second revised edition, New York: Praeger, 1970).*

The Multi-cultural and Multi-national Problems of Armed Forces

RICHARD A. PRESTON

THE UNITED States Army is facing today what some regard as a new problem, difficulties arising from multi-culturalism and multi-nationalism. The former is an aspect of the conflict between black and white currently raging in American civilian society. United States Army policy with regard to black troops was a sorry story until late in World War II. Since 1948, however, when integration was ordered by President Harry S. Truman, the armed forces have been ahead of most of American society in the move to integrate. It is therefore tragic that the pressures of Vietnam have produced new racial tensions and so threatened a promising effort to achieve harmony and to give a lead to the nation.

Multi-national problems for the United States Army stem from the fact that since World War II the country has been increasingly associated with the military of other countries in both war and peace. What seemed during the 1941–1945 conflict to be a highly successful exercise in the art of operating in combination with allies has sometimes been less effective or less happy when the pressures of war are absent; and even in post-1945 war situations there has been friction when cooperation was between peoples whose experiences are very different. Multi-national problems differ from multi-cultural problems because cooperation between allies introduces the factor of inter-governmental relations. This complicates the contact of even those troops who are able to cooperate harmoniously and effectively in professional duties, which is not always easy. In many cases multi-national problems also include cultural conflict, even when allies are closely akin, and much more so when there is great disparity between them. A clash of cultures within alliances may be embittered by differing political goals, objectives, or ambitions.

Multi-cultural and multi-national situations threaten to lower morale and to decrease willing cooperation. They make communications difficult. They thus seem to militate against the capacity of an army to carry out its tasks effectively. They are therefore deserving of special attention. However, these questions have received very little notice by historians. The bibliography of military multi-nationalism and multi-culturalism is surprisingly short and obscure. The result of insufficient historical attention to the subject is that a prevailing concept of military men, that homogeneity and an absolutely undivided command are a *sine qua non* for military success, has barely been tested in the light of history.

That the desirable conditions of homogeneity can be advantageous in armed forces cannot be denied. But the soldier's unwavering belief in them stems from what is a really relatively new concept in the history of human society, namely, that the ideal army is that of a single nation-state. This idea dates from the time of the American and French Revolutions and from the appearance of the ideal of the democratic nation-state. Its relevance to multi-cultural and multi-national problems in armed forces needs to be tested against experience in a wide variety of circumstances over a longer period of time before and after the great democratic revolutions.

An investigation of that magnitude could not be carried out and presented within the scope and space of a paper of this length. If it could be undertaken it would necessarily begin by reminding the reader that history does not give all the answers to current problems. The sum total of history's evidence is not enough for that purpose; selection is inevitably subjective, and circumstances are never exactly repeated. Nevertheless, an examination of past experience of ethno-culturalism in other armed forces may help to throw some light on present American problems and so may serve to help those who have to make policy, even though they are acting now in response to situations that are novel.

The first thing about this problem that strikes the investigator is that the number of examples of multi-nationalism and multi-culturalism in history is very large. Attached as an appendix to this paper is a work list of some well known examples. It does not claim to be either definitive or comprehensive; but a partial and preliminary list of this kind is sufficient to show that military experience in this problem is far from being a new thing, as some have supposed. It has been very much more frequent than has hitherto often been realized. Heterogeneous armies appear in fact to have been the rule rather than the exception. Furthermore, a glance at the list suggests that in most cases ethno-cultural cooperation was successful. At any rate, the number of mixed forces that can be called successful is so large that it immediately casts doubt on the prevailing military theory that homogeneity is a vital essential factor in military success.

In an attempt to analyze the evidence to be derived from the many successful military cooperative efforts listed, which have noticeably occurred in a variety of circumstances, this list is subdivided to group together various types of multi-culturalism and multi-nationalism. By this process it may be possible to isolate factors that have contributed to successful cooperation or have been obstructive to it. Subdivisions range from the cooperation of different ethnic groups or individuals as mercenaries, through the elements in multi-national states where there is undivided allegiance and command, through colonial situations where there is association between an imperial force and colonial troops under established leadership, to ordinary military alliances between nation-states.

THE COMPOSITION OF ARMED FORCES
Mercenaries

The use of aliens as mercenaries dates back to classical times. It has continued to occur throughout history whenever a people prefer to engage in business as usual and can afford to leave its defense to others, or when it feels that it lacks the necessary military expertise to undertake it. Foreign mercenaries have had a bad press, partly because they run counter to the ideal soldier defending his hearth and his home, and partly because they often appear to threaten to undermine the fabric of the democratic state by providing governments with the power to suppress liberty.

Machiavelli, one of the earliest critics of mercenaries in modern times, alleged that they were also militarily ineffective. The *condottieri* of his day employed by Italian city-states had in fact deteriorated from those of earlier times. Fighting was a business for them, and they found it profitable to live to fight and earn on another day. Italian city-state warfare had therefore degenerated (if that is the right word in this context) to a relatively bloodless series of maneuvers in which the prime objective was to seize captives in order to claim ransom and to win victories by movement rather than by blows. By contrasting them with the citizen-soldiers of contemporary European monarchies, Machiavelli found the *condottieri* "useless and dangerous." He wrote: "They are disunited, ambitious, without discipline, faithless, bold among friends, cowardly among enemies, they have no fear of God, and they keep no faith among men." This powerful indictment may have been an inspiration for much of the traditional attitude toward mercenaries after Machiavelli's time. One aspect of it is the use of the word "mercenary" as a pejorative adjective.

But it does not in fact conform with the evidence throughout modern history. As professionals, paid for performance, and trained in their tasks, mercenaries were usually quite effective. In fact, their proficiency was the very quality that was feared by those who thought they endangered liberty. On the whole, they

MULTI-CULTURAL AND MULTI-NATIONAL PROBLEMS

proved to be so efficient that it can be said that they could normally easily overcome obstacles imposed upon their service to their masters by ethno-cultural circumstances. They have also on the whole been faithful to their employers.

Let us take two examples from the list to attempt to find the reason. One of the best known corps of mercenaries in modern times is the French Foreign Legion. It was formed in 1831 by King Louis Philippe to serve in Africa. The Legion was largely responsible for winning and controlling France's foreign possessions in the nineteenth century. Yet it was composed almost exclusively of aliens who had no patriotic loyalty to France. Its success was due to two things: first, the excellence of its training and, second, the effect of its traditions. In the beginning the Legion benefitted from the fact that its original members were veterans from the foreign regiments that had served the restored Bourbons and had been disbanded after the Revolution of 1830. Then and throughout its history the Legion had the inestimable advantage of including large numbers of Germans, always a martial people. Living and training in the desert, out of touch with civilization in Europe, Legionnaires underwent a rigorous training and developed a tradition of loyalty to the Legion and to their comrades that was as effective as patriotism. They did so because they had to in order to survive in conditions of endemic warfare. Professionalism and action thus offset the disadvantages of multi-culturalism.

A second example of the mercenary is the Gurkha who served with the British in India and later with the Indian Army. Although Nepal—whence the Gurkhas came—was in some senses a satellite state, it had retained its independence in all things except relations with foreign powers. Gurkhas were recruited as individuals, as was the case with the Legionnaires, but in their case by virtue of an understanding between the government of British India and the government of Nepal. The export of mercenaries was one of the chief sources of Nepal's income from outside the country. Gurkhas were a kind of military caste and military life was a hereditary tradition with them. The fact that they served a foreign power in no way affected their obedience and their discipline; in fact, it seems to have enhanced it. Gurkhas always lived, worked, and fought in closer harmony with British comrades and British officers than did any Indian group. Discipline and training were the open secret of their success as soldiers. Their performance was not affected by the fact that they were recruited from an alien state by an arrangement sanctioned by an alien authority, for that authority did not attempt to control their use or to tamper with their duty. Gurkhas were professional soldiers first and foremost, and the problems of multi-culturalism did not impede their performance.

Forces in Multi-racial States

The second form of multi-culturalism to be considered is that which occurs in states that have mixed populations. Classical states, and the dynastic states of Europe that preceded the French Revolution, were almost invariably multi-racial in composition. Their armies were therefore often composed of members of different racial groups. Nationalism being then not well developed, racial sentiment offered little obstacle to military service. Dynastic states of this kind frequently employed mercenaries. They also often recruited armies from more than one of the several racial groups of which such states were composed.

Despite the impetus given to nationalism and nationality by the French Revolution, this situation continued into the nineteenth century. Napoleon's army, based in theory on the nationalism of the French, was distinctly multi-national, or perhaps one should say multi-racial, in its composition. It actually included a larger proportion of non-French soldiers than French armies before his time. Dynastic states like Austria-Hungary, the successor to the Holy Roman Empire which Napoleon had destroyed, were much like their predecessors of the eighteenth century. As a result of the growth of armies and European rivalries, and perhaps also of the influence of the nationalist-democratic ideas that prevailed more strongly elsewhere in Europe, monarchs in such states felt impelled to conscript their subjects for military service. Armies were therefore even more multi-racial and polyglot than before the rise of nationalism and democracy.

To some extent this condition could be used to stimulate efficiency by fostering competition. Thus Celtic regiments were found in the British Army; and the Scots paraded with their distinctive dress, the kilt, and their native musical instrument, the bagpipes. The United States, composed of recent immigrants from many different lands, had ethnic regiments made up of Germans and Irish and others during the Civil War. Here again retention of ethnic units was used to foster friendly competition and therefore morale and efficiency. In our own day the Israeli Army, also composed of recruits who have come from many different native lands, is used to develop a common nationalism by indoctrination.

However, multi-racialism in the armies of multi-national states has often had dangerous features. In the Austro-Hungarian Army, it was found necessary to station ethnic regiments far from their own district in order to minimize the possibility of their becoming involved in nationalist uprisings. A second problem was that if an army was recruited disproportionately from one racial group, that group could be put—or seem to be put—in a position to dominate the state because of its grasp on military power. Vice versa, a switch in power from one ethnic group to another might be accompanied by a switch in recruitment in the army, especially in the higher ranks. This seems to have occurred in Belgium, where the officer corps of the Army was at one time almost exclusively Walloon but is now more Flemish. Something of the same kind of thing occurred in South Africa, where the Union's army was originally balanced between the old enemies, Boer and Briton, but where involvement in the "imperial wars" led to a decline of Boer participation at the top. That decline has now given way to a swing in the opposite direction in the army of the Republic.

Truly democratic multi-racial states strive to preserve a balance within their forces similar to the balance in their population. But this is not easy to achieve, especially when some groups have an aversion for military service and others like it. It is also difficult when groups that are underprivileged economically join lower ranks more frequently, while privileged groups supply better educated and better qualified personnel for the officer corps. Such a situation can be inimical to morale and extremely dangerous for the safety of the state. Army training and discipline, normally expected to develop patriotic national spirit, may in fact become a breeding ground for subversive militance.

Where language is a factor, the question of balance is obstructed by problems of communication. Bilingualism, always difficult to attain, often seems to be an impediment to the improvement of communication within the forces, and may be resisted. South Africa did, however, achieve a remarkable degree of bilingualism; but under the influence of strong nationalism it is now said to be moving towards Afrikaans monolingualism. The balance between racial components of armies in bilingual multinational states is thus difficult to maintain. But attempts to achieve it may impair military efficiency.

Colonial Forces

The third group of multi-racial and multi-cultural forces consists of those in colonies and modern overseas empires. European powers expanded throughout the world by employing colonial troops to build—and to help them to administer—their empires. In the process they spawned or developed the idea that certain peoples, usually those who had resisted them most strongly, were better fighting material than others. The so-called "martial races" were on their part glad to get employment as soldiers after the conquest because that was what they had always liked. Expanding empires provided them with plenty of fighting.

This concept or practice of employing martial races was

prevalent in the British Indian Empire. The British developed strong prejudice against those classes or peoples that they considered non-martial. But these were the more civilized of their native subjects, who were collectively described as *"babus,"* that is, clerks. It was among the *babus* that political opposition grew. When the Indian Empire was in its last stages, the martial peoples in the Army remained remarkably faithful until the British flag was lowered. This suggests that military training and discipline had been effective in fostering the loyalty of those native peoples who had experienced it.

After independence the theory of inherent differences between martial and non-martial races became unacceptable in the new India. The Indian Army therefore began to recruit on a national basis. In the early years it actually deliberately recruited disproportionately in an effort to redress the imbalance in the Army. It was possible to do so because of the state of imbalance. The spirit of freedom-born nationalism was said to have made recruiting easier among peoples and classes who had hitherto scorned military service. It is not clear, however, how far this national policy was actually effective. Professor William Gutteridge, the historian of armies in new states, has said categorically, "They are still recruiting from the martial classes."

On the other hand, when the British began to Indianize the officer corps in the years before independence and especially during the Second World War, commissions had gone to men who were adequately educated, that is to say, men from the classes identified as *babus*. The British had always held that Indian troops preferred British officers and would not follow Indian leaders, especially those from a different ethnic origin, and most of all from the despised *babus*. But the upper ranks of the Indian Army, the generals, and brigadiers, are now largely drawn from the educated group rather than from the martial races. The new Indian Army has so far seemed to be effective—for instance, in its struggles with Pakistan. It remains to be seen how far it could still stand the supreme test of a major war if the two countries on the sub-continent resorted to military operations on a considerable scale when more water has flowed under the bridges.

The validity of the prevalent belief in the existence of martial qualities in some peoples and not in others is crucial to the whole study of the question of multi-culturalism in armed forces. Modern ideology tends to deny that such qualities are inherent. It is, however, obvious that social and economic conditions and historical tradition can predispose a man to a soldierly life. On the other hand, it is also possible that military training can "make all men tall." But it takes longer to build an officer corps than to train effective troops in the lower ranks. An important consideration in this respect is technical proficiency or, on the other hand, backwardness. At one time primitive peoples could easily be made into good soldiers, at least in nontechnical or fighting units, and in the lower ranks. In modern warfare this is no longer as true as it once was, and the recruitment of armies now reaches into groups that were once considered nonmilitary. At the same time, service in the army can be a means of providing technical training for primitive peoples, and it is often used for this purpose—for instance, in Turkey. However, it may be that these trends, which suggest that modern technological war might lead to the political emergence of and domination by new groups in new states, are contradicted by what some observers see, namely, a widespread rise of military dictatorship because martial peoples trained by the old imperial powers are taking over in very many places.

Alliances

The fourth group of examples of multi-culturalism and multi-national forces to be considered are alliances of independent nation-states. This group differs from all those discussed earlier because armies in alliance serve different political masters. The degree of possible political interference or influence is therefore greater than in all the other groups discussed above.

MULTI-CULTURAL AND MULTI-NATIONAL PROBLEMS

Alliance between advanced states means cooperation between armed forces that have already developed their own organization and communications systems. Compromises to foster cooperation are not easy to be achieved in face of military conservatism. Language is also likely to be more persistent as a barrier to effective communication unless the allies happen to speak the same tongue.

Thorough cooperation between an advanced power and smaller or backward allies is in some ways even more difficult to operate. There is always a tendency for the strong power to seek to "go it alone" or to insist on full control. This is partly because the aid of small allies, apart from a certain propaganda value, may not be worth the effort that is required to develop it fully. Such allies may be permitted only minor roles, as service troops or as combat forces on quiet fronts.

But the restriction of smaller allies to inconsequential roles is also due to a factor that is often overlooked. Each ally has different interests and objectives. These prescribe the nature of its contribution and the war policy that it seeks to pursue. The stronger power is therefore willing to shoulder greater responsibility in order to ensure that its own interests are achieved and that those of its allies are not.

Problems of this kind govern such questions as appointment to combined commands and the degree of integration of forces. The morale of elements of combined forces can be fostered by national cohesion in units of the largest possible size; and it is in the interest of a power to have its forces intact and under its own commanders when it sits down at the peace conference table. But for military effectiveness, if communication problems can be overcome, the best form of integration would seem to be that each power should contribute the kind of forces that it can most readily supply. However, a smaller power thereby surrenders its capacity to operate independently. Although it might be argued that if a small power contributes a vital arm or service to a combined enterprise the possibility of a threat of withdrawal might give it even more leverage in disputes with its ally, the larger power can probably afford to overlook such a threat because it can usually make up the deficiency from its own resources.

Successful cooperation of allies must thus overcome very great obstacles. It needs a high quality of personal leadership at the top, of the kind provided by Marlborough in the Grand Alliance against Louis XIV, or by Eisenhower in some parts of the later "Grand Alliance" against Hitler. It also necessitates forbearance and understanding at lower levels. It is expedited if there is an earlier tradition of cooperation between the allies, as there was among the great self-governing British dominions which made successful contributions to the British effort in the First World War and again in the Second. It is noticeable, however, that although Commonwealth cooperation might perhaps be described as possibly the most successful example of the cooperation of allied forces in war, achieving far more integration in some ways than that attained by almost any other group of allies, dominion leaders were fully aware of the importance of retaining their military individuality and of the need to build it up rather than let it be surrendered because of the need to win the war.

Some General Conclusions

We have seen that military training and discipline can overcome many of the problems that arise from multi-national and multi-cultural conflict within armed forces. We have also seen that the effect of training and discipline can be turned aside when political influences intervene. In multi-racial states, and to a lesser extent in colonial armies, political influences from civil society can, in certain circumstances, jeopardize military effectiveness. As is usual with military force, pressures of this kind become stronger when armies suffer reverses. In democratic societies and societies where constitutionalism prevails, diverse factors in civilian life will eventually work their way into armies that are raised by conscription and will undermine them. But the replacement of conscription by a professional army may con-

MULTI-CULTURAL AND MULTI-NATIONAL PROBLEMS

tinue to threaten the security of that state if the army is recruited from disadvantaged elements. Relations with allies can also be disturbed by policies that pursue selfish national interests.

Yet we have also seen that in the course of history, multi-cultural and multi-national forces have been far more successful than is usually realized. Despite the seeming disadvantages of heterogeneity, it must somehow generate powerful constructive forces that can be exploited for the general good provided policy is fashioned for that purpose and not for particularist interests. The wounds caused by division within the army cannot be healed unless the wounds in society are also treated. Good leadership, civilian and military, can go far to achieve this end. Within the armed forces healthy indoctrination and the encouragement of sound national sentiment and of reasonable cultural distinctiveness is a necessary accompaniment to normal military training and discipline. Armed forces cannot reform nations merely by their example, but their integrity is essential if nations are to survive. The world appears to be moving towards cultural pluralism within a global context, and armed forces must move with it. This seems to be the lesson that the history of multi-culturalism and multi-nationalism in armed forces has to teach.

Appendix
Multi-cultural Military Forces—Examples

A. Multi-cultural and Multi-ethnic Forces
  1. The Employment of Foreigners.
     Mercenaries in Classical, Medieval and Modern States.
       William the Conqueror's force at Hastings.
       The armies of Carthage.
       Genoese crossbowmen, German Ritters, Swiss mercenaries.
       The Janissaries.
       Italian *condottieri*.
       Prussian troops recruited throughout Europe.
       Hessians and Hanoverians in the pay of George III.
       Cossacks in imperial Russian armies.
       Gurkhas in British and Indian pay.
  2. Multi-racial states.
     a. Ancient Empires.
        Alexander the Great's army.
        Legions of the Roman Empire.
     b. Volunteers.
        Germans and French Canadians recruited for the Union Army in the American Civil War. Scots, Irish, and English Jacobites in the pay of the King of France.
        The International Brigade in the Spanish Civil War.
        American volunteers in the RAF "Eagle" Squadron.
     c. Modern Polyglot Empires.
        Napoleon's Grand Army.
        Austria-Hungary.
        The Russian Empire.
     d. Modern Republics and Constitutional Monarchies.
        Irish regiments in the Union Army in the Civil War.
        French, German, Italian, and Romance cantons in the Swiss Army.
        Irish and Scottish regiments in the British Army.
        Flemish and Walloons in the Belgian Army.
        Ethnic groups in the Soviet Army.
        French Canadian units.
        English-speaking and Afrikaans-speaking South Africans.
        The Israeli Army.
        American Negro troops in the Civil War, in the West, and in

MULTI-CULTURAL AND MULTI-NATIONAL PROBLEMS

   two world wars.
   Japanese Nisei in the Second World War.
  e. Modern Colonial Empires.
   The British Indian Army.
   British African regiments.
   The Arab Legion in British pay.
   French Colonial troops, African and Asian.
   Indian Scouts.
   Filipino troops.
   Spanish Moors.
   "Native" troops in South Africa.
B. Multi-national Forces
 1. Alliances.
   Greek city-states.
   Marlborough's Grand Alliance army and the Earl of Galway's force in the Peninsula.
   French allies during the American Revolution.
   Satellite kingdom troops in Napoleon's army.
   Wellington's Peninsular army.
   The European allies on the continent up to and including Waterloo.
   The British, French, and Piedmontese in the Crimea.
   The Allies in the First World War.
   The Central Powers in the First World War.
   Allenby's Arab allies.
   The United Nations in the Second World War.
   The Axis.
   Australasian and Asian allies of the United States in Vietnam.
   The North Atlantic Treaty Organization.
 2. Pseudovolunteers.
   Elizabethan forces in the Netherlands.
   Italian "volunteers" in the Spanish Civil War.
   Chinese "volunteers" in Korea.
 3. The British Empire and Commonwealth.
   Indian princely-state armies.
   Canadian, Anzac, South African, Indian, and other troops in two world wars, *e.g.,* the Eighth Army in Egypt.
   The Commonwealth Division in Korea.
 4. International.
   The Crusades.
   "United Nations" Forces in Korea.
   Peacekeeping forces in the Middle East, Congo, Cyprus, etc.

COMMENTS

According to the customary generalizations already alluded to, about armies as reflections of their societies, the Austro-Hungarian Army should hardly have functioned at all in the Great War of 1914–1918. It was the army of a decrepit empire. It was also the army of a society utterly lacking in ethnic and cultural cohesion. Yet what is striking about the Habsburg army of the Great War, considering the state and the society from which it sprang, is not its defects but its toughness and cohesion. Of course, it was not the German Army; its weaknesses were glaring compared with the army of Austria's principal ally. But the Habsburg army held together through four years of combat in the climatic extremes of the Eastern Front and then among the mountains of the Austro-Italian frontier, and in the end the army collapsed only after the disintegration not merely of the society but of the very government it served. Let the reader reflect on the vitality and endurance that an army can generate for itself. Gunther E. Rothenberg of Purdue University can probably give a sounder introduction to the Austro-Hungarian Army than any other living military historian. He has published **The Military Border in Croatia, 1740–1881** *(Chicago: University of Chicago Press, 1966) and will soon publish a general history of the Habsburg army. He is himself a veteran of three armed forces—the British Army, the Israeli Army, and the United States Air Force.*

The Army of Austria-Hungary, 1868-1918: A Case Study of a Multi-ethnic Force

GUNTHER E. ROTHENBERG

FOLLOWING the striking Prussian victories of 1864, 1866, and 1870, all of which were achieved by a conscript army drawn from the entire population, the principle of the nation in arms gained rapid and universal acceptance in continental Europe. To be sure, total conscription was an ideal, not usually achieved in reality, and none of the European powers operated this system to its fullest capacity. Even so, conscription tended to bring into the army all elements of the population, and it often was regarded by its advocates as the "school of the nation," bringing to recruits from all walks of life a greater sense of national unity and purpose.

The issue was complicated by the fact that the nation-state too was only an ideal. Many European nation-states were not nationally homogeneous and continued to have minority-group elements in their armies. This was true of the Alsatians and Poles in the German Army, of Italians in the French Army, and of Poles, Ukrainians, Jews, and others in the Russian Army. The most complex case, however, was that of the Imperial and Royal Army, the army of Austria-Hungary, which, according to official count, numbered 267 Germans, 223 Hungarians, 135 Czechs, 38 Slovaks, 85 Poles, 81 Ruthenes, 26 Slovenes, 67 Croats and Serbs, 64 Rumanians, and 14 Italians for every 1,000 men in its ranks. Here, then, there was truly an army of a multi-

THE COMPOSITION OF ARMED FORCES

national state; and to the degree that in the nineteenth century national aspirations became more pronounced and more contradictory, they were reflected in the army.

The apparently inherent weakness of such an army was well recognized by its leaders and by foreign observers. Already in 1855 Friedrich Engels commented that "no one can predict how long this army will stay together," and in 1895 Count Casimir Badeni, Austria's Prime Minister, remarked that "a state of nationalities can make no war without danger to itself." Such pessimism also was expressed in the highest command echelons. In November 1906 Feldmarschall Leutnant Franz Conrad von Hötzendorf, Chief of the General Staff for the "entire armed forces" of the Austro-Hungarian monarchy, and an untiring advocate of preemptive war, argued that "all preparations for foreign war are useless as long as our internal situation is not resolved." Specifically, Conrad referred here to the continuing conflict over the composition and the character of the military establishment which had bitterly divided the empire since 1867 and which, in 1905, had brought it to the very brink of civil war.

The problems inherent in a multi-national army based on conscription had long been familiar to the rulers and counselors of the Hapsburg monarchy, which, since its inception in the early sixteenth century, had always been a multi-national state in which the various parts had always been able, or tried, to assert a large degree of autonomy. The army had always been regarded as an internal as well as an external support of the monarchy. The typical dynastic army, which emerged in the years after the seventeenth century, an officer corps recruited from the nobility and loyal to the sovereign, and a rank and file drawn from the lower classes, without political consciousness and strictly regimented, had served such purposes adequately. Even when, during the era of the French Revolution and Napoleon, other continental powers reluctantly adopted conscription to fight the French mass armies, the Hapsburgs retained their professional force. Conscription, they believed, undermined the position of absolute monarchs, and at the Congress of Vienna, Prince Metternich, their chief minister, championed the abolition of conscription. In the period up to 1866, therefore, Austria essentially maintained an eighteenth-century military establishment in which, moreover, units were commonly stationed outside their ethnic areas in order to isolate them from potentially subversive influences.

Of course, the system had its drawbacks. Nationalist and subversive influences could not be excluded entirely, and the professional army could not mobilize the full economic and social potential of the empire. Moreover, the dispersal of troops made mobilization proper a slow and cumbersome business. After being defeated by Prussian conscript armies in 1866, Emperor Francis Joseph reluctantly decided to introduce a system of universal short-service conscription—which resulted in the multi-national army whose composition was given above. However, this action aroused opposition. On the one hand stood the imperial generalcy, headed by Archduke Albrecht, who maintained that only a professional, dynastically oriented army could possibly hold the monarchy together; on the other hand stood the Hungarian nationalists, who for many years had aspired to their own national army, and who after the debacle of 1866 were powerful enough to obtain at least part of their demands.

Politically, the Hungarian demands were met in the Compromise of 1867, which created the Dual Monarchy of Austria-Hungary and gave the Hungarians political parity. The military question was settled in a separate military compromise in 1868. Originally, Hungary had demanded an independent national army, but now it accepted the continuation of the imperial-royal army, also referred to as the "joint" army, and the creation of two separate auxiliary forces, the *Honvéd* and the *Landwehr*, in the two halves of the monarchy. All male subjects of the Dual Monarchy were liable for military service, and a portion of the annual recruit intake, the exact quota to be fixed by decennial negotiations between Vienna and Budapest, went directly into the auxiliary forces. Moreover, while the joint army remained under the exclusive control of the Emperor-King and his appointees, *Honvéd* and *Landwehr* were to be controlled by national defense ministries in Budapest and Vienna and came under central command only in times of war.

From the outset the settlement was not entirely satisfactory to either side, and one historian recently has called it "the Achilles heel of the dualistic system." Because of the generous support provided by the Hungarian Parliament, the *Honvéd*, using Hungarian insignia and language, and at least partially the realization of a national dream, soon became a very considerable force; but it still remained a second-line organization. Moreover, the continued presence of joint army units, the most visible and vital instrument of imperial unity, was resented by national extremists. For some twenty years, nevertheless, the military compromise muted differences; but from 1888 on the Hungarians began to use their political leverage to gain greater autonomy for the *Honvéd* and increased influence in the joint army. Their ultimate goal was a completely independent national Hungarian "language of command" in all units of the joint army recruited on Hungarian territory, regardless of whether they were ethnically Hungarian or not.

The question of the "language of command" moved into the forefront of the struggle regarding the joint army. This language consisted of some eighty drill phrases, learned mainly by rote, and their retention was regarded by the Emperor and his generals as absolutely essential, not only as a means of communication but also as the symbol of the continuing unity of the army. The joint army also recognized ten different "regimental languages," used for the day-to-day affairs and for instruction within a unit. Again, there was little common ground here. In 1906, out of 256 units in the joint army, 94 used but one language, 133 used two, while 28 authorized the use of three and even four regimental languages. To control the language of command as well as the language of instruction would give a nationality a considerable advantage in imposing its will. For instance, the Magyars in Hungary, a minority in the overall population of the kingdom, regarded the imposition of their language as an absolutely necessary step towards absorbing their national minorities. For precisely this reason the joint army, which stood both "in principle and in reality on the basis of national equality," opposed the Hungarian demands. As Conrad put it: "The unity of the army, which is the vital prerequisite for the continued existence of the monarchy, can only be maintained by guaranteeing equal treatment for all nationalities."

This then was the basic issue between Vienna and Budapest. Of course, under the settlement of 1867-1868 the Budapest government had little influence over the internal affairs of the joint army, but it could make its weight felt during the decennial renegotiations of the army budget. Hungary demanded concessions as the price of its assent for expansion and rearmament of the joint army and, failing to get such concessions, tied up passage of the measures. Parliamentary obstruction, periodically supported by public demonstrations and riots, became a serious matter for Vienna. By 1905 the situation was regarded as so critical that plans to occupy Hungary with military forces composed of German, Croatian, Czech, and Rumanian units were being prepared in Vienna. At the last moment, however, Emperor Francis Joseph refused to sanction this drastic action and instead made the necessary concessions to Hungary.

By 1912 the struggle over the Hungarian demands was, at least temporarily, over. The *Honvéd* now approached in all essentials the characteristics of a national army, though, when war came the joint army and the *Honvéd,* for all their spiritual incompatibility, fought well together. On the other hand, the struggle had delayed expansion of the joint army, and during a period when all European states were arming feverishly, the Austro-Hungarian military budget remained the lowest among the great powers. Although the population of the empire rose over 40 percent between 1870 and 1914, the military establishment increased only 12 percent, and the absent reserves and the outdated equipment would be badly missed in 1914-1915. As an English writer recently put it: "The weakness of the Hapsburg army in 1914 stemmed not from the disaffection of its soldiers but from the intransigence of politicians in Hungary."

This summation is basically correct, but it does require

THE COMPOSITION OF ARMED FORCES

elaboration. Adoption of universal military service required that the masses of the people be kept fairly contented, but the political and military advantages conceded to the Hungarians in 1867-1868 had fanned the antagonisms and aspirations of other national groups in the Hapsburg Empire. Though the degree of intensity varied, they too were now making demands for national influence within the military establishment. The South Slavs, for instance, were up in arms, and attempts to enforce conscription provoked two bloody uprisings in Dalmatia in 1869 and 1881; and there also was armed opposition to the occupation regime in Bosnia-Hercegovina. By the 1890s the Poles were asking for a share in the control of the *Landwehr*, and even the Germans, seeing their former primacy challenged, were becoming restive. Above all there was trouble in Bohemia, where, by the turn of the century, the struggle for national and ethnic assertion combined with social agitation. Certain units proved unreliable for riot duty, and during the annual reserve musters of 1905-1906 Czech reservists refused to answer roll call. More ominous yet, during the Balkan crisis of 1912-1913 a substantial number of Czech reservists refused to heed mobilization notices, and one unit mutinied when it was transported to the Serb frontier. The army, it appeared, was becoming a blunted instrument for war and unsuitable for suppressing popular revolts. Both Conrad and Archduke Francis Ferdinand, heir to the throne and army inspector general, were so concerned about the reliability of the troops that they were willing to compromise rapid mobilization in favor of the old expedient of shifting troops out of their home districts.

Not all observers, however, felt that the Austro-Hungarian Army was in such bad shape. In 1913 an English critic of the monarchy wrote that in the midst of the nationality struggle the "Army is a school of unitary sentiment and a constant corrective to particularist ambitions." He admitted that the tripartite military organization tended to be cumbersome, but he asserted that it worked. "The maintenance of unitary sentiment and of efficient organization in this maze of languages and races," he con-

A MULTI-ETHNIC FORCE

cluded, "is a dynastic and military miracle—a miracle accomplished by the devotion of its corps of officers." Similar conclusions were reached in a German General Staff memorandum of the same year which declared that the Hapsburg officer corps still formed the "main and at this time quite effective counterbalance to the polyglot character of the army."

This "military miracle" was accomplished by a very small officer corps, some 17,500 active-duty regular officers, assisted by some 12,000 reserve officers, in a peacetime establishment of over 450,000 men. Both the composition and the character of the corps had changed since 1868. As a whole, the Austrian officer always had been known for his bravery, dash, and devotion to the imperial house. On the other hand, he rarely had been a learned man. Upper ranks often had been filled by aristocratic amateurs; the lower ranks had stressed individual courage and a dogged adherence to the drill manual. Overall this had resulted in poor staff work and a cavalier disregard for technological innovations. But after 1868 the new requirements of a mass army—new weapons, tactics, and communications—required much greater professional expertise, and military careers in Austria-Hungary became open to talent, requiring "brains, aptitude for work and fitness for command."

The officer corps retained a predominantly "German" character. By the last decade before the war, although Germans comprised only 24 percent of the total population and 25 percent of all ranks in the armed forces, they still provided 70 percent of all regular officers. National background, however, was no bar to entry or promotion, and for many Czechs, Croats, or Poles the army offered a very tempting career. National origin was no bar to promotion, and in the highest echelons some 16 percent of all general officers were of Croat nationality. Religion also was no hindrance. While Roman Catholics predominated in the corps, 86.5 percent among the regulars, there also were Protestant, Orthodox, and Jewish regular officers. In fact, some German writers reproached the corps for being contaminated with "Jews, democrats, and freethinkers." Other foreign observers, however, were much impressed with the Austro-Hungarian officers, and one English writer rated them superior to the average German officer. The Austro-Hungarian officer, he wrote, "is more intelligent, more readily adaptable to circumstances, in closer touch with his men, less given to dissipation, and remarkably free from arrogance."

This was too favorable an evaluation, but the regular Austro-Hungarian officer, staff as well as line, was a hardworking man. The composition of the army forced him to know several languages, and especially in the line units there were good relations between officers and men. At that, conditions for the enlisted men—pay, rations, and quarters—were not good; Austria-Hungary spent somewhat less per year on each man than France and much less than Germany. Because of a continued shortage of senior noncommissioned officers, many duties, which in other armies normally were handled by sergeants, had to be handled by junior officers; and this may have contributed to closer contacts among all ranks. At any rate, the Austro-Hungarian officer was able to make effective troops out of unpromising recruits. For instance, the tough Bosnians who fought against Austria in 1878 and 1881, and continued in a state of endemic insurgency for several years thereafter, provided by the 1890s regiments which were regarded as highly reliable and justified this reputation throughout the war.

The strong dynastic sentiment and the dedication of the professionals was not always shared by the reserve officers whose services were needed after the introduction of universal military service. These officers, recruited from the ranks of the "one-year volunteers," usually came from national groups with a considerable and educated middle class. Germans provided 60 percent, Magyars 24 percent, and Czechs 10 percent, while Poles, Ukrainians, Slovenes, and Rumanians provided only insignificant numbers. In Austria-Hungary, moreover, a reserve commission never held the prestige it did in Germany, and normally, reserve officers had neither the time nor the inclination to develop an *esprit de corps* or enter into the closed circle of the

A MULTI-ETHNIC FORCE

professionals. They usually retained strong ties to the national and social concerns of their background and in consequence were often looked on with suspicion by the regulars.

Nevertheless, it was the reserves who together with the regulars absorbed the initial shock of war, and contrary to often-stated opinion, the Hapsburg army proved itself an effective fighting force in 1914–1918. Much to the surprise of Austria's enemies and to the relief of the imperial and royal command, the initial mobilization went without a hitch. Fears regarding *Honvéd* proved without foundation, and there existed, even among Czechs and Serbs, unsuspected reservoirs of dynastic loyalty. Yet there were weaknesses. From the outset, fears of national disaffection, real or imagined, limited the deployment of various national elements. In general, only German, Magyar, and Bosnian regiments could be committed to battle on all fronts. And after the initial enthusiasm had worn off, the combat value of other nationalities varied. The first serious signs of trouble came in Bohemia, where as early as September, 1914, the departure of some reserve formations to the front provoked antiwar demonstrations, followed in April, 1915, by the mass defection of one unit, the 28th Infantry from Prague, to the Russians.

As the war continued, disaffection appeared among other national contingents, a trend facilitated by the severe losses in professional officers and cadres sustained at the start of the war. "The old professional army," the official Austrian history states, "died in 1914." But the army still had recuperative powers. The losses were made good, and by December the Austro-Hungarian Army was already able to take the offensive once again, though officer replacements remained a problem for the rest of the war and the cohesion of many units suffered.

After the death of the old Emperor Francis Joseph in 1916, and with the tide of war turning against the monarchy in the winter of 1917, there were increasing numbers of incidents involving not only Czech, but also Serb, Ruthene, Slovene, and even Croat units. Here the sense of solidarity with their ethnic

THE COMPOSITION OF ARMED FORCES

group began to gain over fidelity to the Emperor and the monarchy. By the winter of 1916-1917 many regiments had to be considered as unreliable and units had to be mixed up or brigaded with others of a different nationality in order to ensure their conduct. Even so, much depended on individual commanders. In 1917, for instance, the 99th Infantry, predominantly Czech and from a region where disaffection was widespread, distinguished itself during the battle of Caporetto.

In fact, the problem of disaffection within the armed forces of the Hapsburg monarchy must be seen in perspective. The German Army, a much more homogeneous force, also had considerable numbers of Polish and Alsatian deserters, and no large-scale mutinies like those in the French and Russian Armies took place in the Austrian Army. To be sure, there were storm signals. A great wave of industrial strikes in January, 1918 was followed by the mutiny of the Fifth Fleet in the Gulf of Kotor in February, both demonstrations against poor rations and for an early peace. Even so, both were broken by the army, and until the early summer of that year the Austro-Hungarian forces held their lines and in June, 1918 mounted their last great offensive against Italy. The failure of this offensive, together with the influence of the Bolshevik Revolution in Russia, and the emergence of national leaders who no longer looked for changes within the monarchy but were committed to its total disruption, broke the cohesion of the weary and hungry troops. After May, 1918, mutinies, refusals to go into action, and mass desertions became frequent. In some areas the "green cadres," armed groups of deserters, began to constitute nuclei of armed resistance movements. By the late summer of 1918 the army, though it still stubbornly defended itself in Italy and was conducting a fighting retreat in the Balkans, was clearly near the end of its rope.

The final blow, however, was the imperial manifesto of October 16, 1918, in which the Emperor Charles conceded to his nationalities the right to form their own states. It was followed within two weeks by an imperial order releasing officers from

A MULTI-ETHNIC FORCE

their allegiance and permitting them to join the national armies of the new states. These developments destroyed the last bonds keeping the army together. On October 29-30, 1918, the new national governments, Hungarian, Czecho-Slovak, and Croatian (Jugoslav), recalled their troops, and the holes this left in the front could no longer be plugged. In any case, already on October 24, 1918, an Italian offensive, spearheaded by British and French troops, had made considerable progress; but still the final breakdown came in the capitals of the empire and not at the the front.

Seen in overall perspective this multinational force, underequipped and outnumbered, fought extremely well. While the aspects of national resistance have been stressed by historians of the successor states, B.H. Liddell Hart concluded that "this loosely knit conglomeration of races withstood the shock and strain of war for four years in a way that surprised and dismayed her opponents."

Tradition, discipline, and dedicated leadership were the mainstays of the Hapsburg army. It recently has been suggested that other solutions to the military problem of the Dual Monarchy could have been found after 1866. It is argued that there was nothing essential in the structure of the monarchy to compel it to adopt conscription and that a truly professional army would have served its interests better. On the other hand, a prominent Austrian military historian suggested that the military compromise of 1868 was only a reluctant accommodation to changes imposed by defeat and that it failed to revitalize the military structure. Continued reliance on tradition and on dynastic sentiment, Kurt Peball argues, prevented the army from gaining popular support. He suggests that a truly "multinational army," on a federal basis, would perhaps have been the answer.

It is always fascinating to try to rewrite history and to undo with a pen what has been done with the sword. The unexpected performance of the Hapsburg army during the First World War was in large part based upon a feeling that the government was legitimate and was pursuing a goal acceptable to the vast majority of the multinational population. It was the inconclusive nature of the war, and after 1916 the death of the old Emperor and the obvious tilting of the balance against Austria-Hungary and her allies, that eroded the legitimacy of the Hapsburg government in the eyes of various subject nationalities. To prevent this process would have required a political as well as a military answer.

Militarily, of course, if the war had been won by 1916, there would have been no problem. The suggestions made that the army could have been reconstructed on either a purely professional or a truly multinational base fail to see that these would have been political and not military answers. By 1868 the tide of liberalism was strongly opposed to the continuation of the old-fashioned eighteenth-century army, while at the same time the struggle of the nationalities doomed prospects for a truly multinational federated force. In such a case, it would have been necessary to ensure the loyalty of the various nationalities and classes by extending to them various rights. In Austria-Hungary the adoption of universal service coincided with political reforms through which the monarchy became constitutional instead of absolute; but these reforms did not go very far in Hungary, where the Magyar minority was quite unwilling to share power with the national minorities, and the Emperor shrank from imposing his will at the cost of civil war. At that, as the struggle between the nationalities assumed ever more virulent forms, neither the Germans nor the Slavs of the monarchy would have accepted such a "federated military" establishment. The fact remains that to the degree that the nationality problem could not be resolved in the Hapsburg monarchy, or for that matter in Central Europe, it also could not be resolved within the army. The Hapsburg army could not save the society into which it was placed, but it remains to its credit that it tried.

V

Armed Forces in Politics and Diplomacy

The Petrograd Garrison and
the February Revolution
of 1917

WARREN B. WALSH

page 256

Multilateral Intervention
in Russia, 1918–1919:
Three Levels of Complexity

GADDIS SMITH

page 274

Military Government:
Two Approaches,
Russian and American

EARL P. ZIEMKE

page 290

COMMENTS

The fears of military intervention in politics that led the writers of the United States Constitution so carefully to hedge about military power with checks and balances were fears of an outright military coup d'état—*bayonets dispersing a parliamentary assembly, as the founding fathers well knew had occurred in England the previous century, and had seemed threatened in America at the time of the recent Newburgh Addresses. Military intervention in politics today may still take the form of a* coup d'état, *but almost always when it does so the scene is an underdeveloped, unindustrialized state in which the military is one of the few cadres of educated, energetic leadership and perhaps the only one with a national vision. In the great powers, military intervention and influence are likely to be much more complex and sophisticated affairs than a* coup d'état.

But power is still so close to the essence of politics that the possibility of military power's baring its bayonets can never be totally absent even in the most sophisticated state where the army is strong. In the most severe of national crises, no state can count on being altogether different from the Russia of 1917–where, as Warren B. Walsh described in his 1971 lecture, the once elite and disciplined Petrograd garrison was torn by the national disintegration surrounding it into the channels of revolution. Professor Walsh of Syracuse University has long studied the soldier and the state in Russia.

The Petrograd Garrison and the February Revolution of 1917

WARREN B. WALSH

> Finally, I said to him [General M. V. Alexeev, Chief of Staff, Supreme Headquarters of the Russian Armies]: 'Isn't it strange? One might say that our roles are inverted. You, Aide-de-Camp General to the Emperor, a member of his entourage, you are opposing the Monarchy. And I, a member of the opposition [Maklakov was a leader of the Constitutional Democratic Party in the State Duma], I beg for it!
>
> 'You are correct,' responded the general: 'it is precisely because I know the true Monarchy better than you that I want no part of it.' This observation struck me. 'It is only possible for me,' I replied in turn, 'because I know the politicians better than you, and that is why I expect nothing from your venture.'
>
> V.A. Maklakov in *La chute du régime tsariste*

THIS EXCHANGE between the liberal political leader and the onetime Chief of Staff at Stavka took place some six months after the events with which this paper is concerned, when General Alexeev consulted Maklakov about a plan first to destroy the Bolsheviks and other revolutionaries and then to replace disorder with "a revolutionary order." That sounds confusing because it was confusing, and the General, along with many other responsible people, was both baffled and confused. What happened in the Russian Empire in February/March, 1917, and for months thereafter was, in the words of Milton's *Paradise*

Lost, "With ruin upon ruin, rout upon rout/Confusion worse confounded."

The February/March Revolution throughout the Russian Empire was triggered by what happened and was continuing to happen at Petrograd and its environs and at Mogiliev, seat of the Supreme Command of the Russian Armies (STAVKA), about 425 miles south of the capital. Emperor Nicholas II had left his capital on 22 February*, reaching Mogilieve the next day. In Petrograd there were already strikes, street demonstrations, and disorders, the latter mostly the work of hooligans, before the emperor left for STAVKA. His absence from the capital and his presence at STAVKA certainly made some difference in the happenings at both places, but these were not of sufficient moment to qualify as events except for which subsequent happenings would probably have been significantly different. To argue otherwise one must assume that Nicholas would suddenly have been able to do what he had never done before: take effective charge in a crisis.

The spread of the revolution from Petrograd and STAVKA first to Moscow and later to other civilian and military centers lies beyond the scope of this paper. It may be mentioned as a matter of interest that the first quasi-official news of events in Petrograd was a telegram sent by M.V. Rodzianko, who signed it as President of the State Duma, and by a Duma member named Bublikov, who had taken charge of the Ministry of the Ways of Communication on the preceding day. The telegram went to all railway stations and to all army telegraph units throughout the empire. It stated, somewhat prematurely, that the State Duma was forming a new regime because the old one, having ruined everything, had lost all power. Lieutenant General N.N. Golovin, one of the few soldier-scholars of the Imperial Russian Army, wrote in 1920 that this Bublikov-Rodzianko telegram did more to destroy the Army than the more notorious Order No. One. Golovin had been Chief of Staff on the South-

---

*The calendar used in Russia through 1917 was 13 days behind the western calendar. The Russian "old style" calendar is used throughout this paper.

THE FEBRUARY REVOLUTION OF 1917

western Front when the telegram arrived. He somehow failed to spot it and sequester it, so that news of its content soon leaked to the troops, who concluded from this that the Tsar had fallen. It is difficult to understand the impact of this telegram and of later news unless one has in mind a few basic facts about the tsarist system of government and administration.

Obviously, no man, even if he considered himself as endowed by Almighty God with absolute power over and responsibility for Russia, could rule so vast an empire singlehandedly. On the one hand, a Tsar was utterly dependent upon those whom he considered as his servants, which is how Nicholas II, like his predecessors, thought of even the highest military officers and civilian officials. They were his eyes and ears, his communicators, and his hands. They were also the buffers between a Tsar and all but a few hundred of his subjects. They reported to him, took his orders, and implemented them in varying degrees and with varying efficiency. If a Tsar could not function without the military, especially the army, and the imperial bureaucracy, they could not function without him; this is "the other hand." Power stemmed solely, wholly from the Tsar. A minister did not have power because he held a ministerial post; the power did not, in other words, attach to the position. A minister acted for and in the name of and by the authority of the Tsar. The same was true of the highest-ranking military officer. You are all familiar with a device called a sole-source contract, to be used supposedly when only one supplier can be counted on to produce whatever is required. The Tsar was, so to speak, the sole-source contractor for power. There were instances when this principle was violated in practice, just as there perhaps have been abuses of the sole-source contract. But by and large, if you can think of the Imperial Russian government and administration as analogous to an electric power system, the Tsar was the sole source producing energy for the system. If the Tsar-generator failed, or if components of the system malfunctioned—and both things were happening before 1917 in Russia—, the system faltered and, in the end, failed.

ARMED FORCES IN POLITICS AND DIPLOMACY

Civilians find it as easy to assume that an army is unchanging and institutionally immortal as to assume the existence of a "military mind." The Russian Army underwent at least three major changes during the last sixty-odd years of the Empire. The military reforms of the 1860s were a major watershed, so that one may properly speak of a pre-reform and a post-reform army. This is also true of what happened following the defeat of Russia by Japan in 1904-1905. The Russian Army that began World War I was different in many respects from the shattered forces that returned from East Asia in 1905-1906. For one thing, the proportion of literate recruits in 1913 had reached the unprecedented figure of 68 percent. A military audience certainly needs no reminder that methods of training, indoctrination, and command that are indispensable in dealing with illiterates can prove counterproductive when applied to literate and better educated personnel.

To follow the same line further for a moment, the tremendous requirements for manpower during World War I brought into the army a larger leaven of a new type of men as well as of officers. At the outset of the war a new regulation authorized the immediate promotion, without the examinations that had previously been required, of all volunteers and recruits who had completed at least three-quarters of their secondary education. Those with technical education were offered not only promotion, but also assignment to noncombat duty in the fields of their technical competence. These devices were obviously designed to meet the emergency demands for junior officers, but such devices could be used only after an increase in literacy and education. Moreover, the national emergency brought into the Army not only eager young cadets but also older civilians who had already made their individual marks in the professions or elsewhere. They brought with them different values, different aspirations and ambitions. The civilian trick, accepted if not abetted by the Army, of getting rid of "trouble-makers" by committing them to military service introduced still another sort of leaven into the loaf. The loaf remained substantially peasant,

THE FEBRUARY REVOLUTION OF 1917

as we shall see later, but the new leaven included men, some of whom were successful political activists and agitators, as well as others who had attained recognition as leaders among their civilian peers.

Changes had also taken place, as hinted above, in the officer corps, even in the highest echelons. Such changes had begun with the rebuilding of the Army after 1906 and had been slowly but steadily accelerated even before 1914. This is an important and fascinating story that needs telling, but not at this moment. It must suffice for now to state simply that the special relationship which had existed for generations between the reigning Emperor and all ranks of the military, but especially in the officer corps, began to change after 1905-1906. The almost mystical adoration of the reigning Emperor persisted among a majority but not among an ever-growing minority. The outbreak of war in 1914 revived the old feeling, and there was a tremendous emotional surge "for God, for the Emperor, and for Russia," to quote the words of the traditional oath. Such emotions persisted to the end for some, but in decreasing force and with increasing exceptions to what had been the general rule. We shall return to this point after a look in two other directions.

Successive rulers of Russia from its beginnings had regularly assigned to their military general functions and specific duties far removed from warring against foreign enemies. The famous *streltsi* of the sixteenth century began as bodyguards for the rulers and as security police for the capital, but they also served as Moscow's fire fighters. The rulers depended heavily upon army officers as advisers, ministers, special envoys, political administrators, and general troubleshooters. This practice was slightly reduced but never abandoned after an imperial civil service was developed. Troops were regularly used to protect the imperial court and the person of the sovereign; but they were also used to collect taxes, to quarantine plague zones, and to maintain civil order. Certain patterns developed and persisted. Special care was taken to keep all off-duty military, both officers and men, from forming close connections with civilians. Great

Russian soldiers were assigned to Poland and to the border regions. Poles, Bielorussians, and Ukrainians were sent to the Urals and the Caucasus. Special responsibilities were given to the elite Guards Regiments, which were almost entirely officered by noblemen even into the twentieth century.

It was taken for granted that the military, especially the officer corps and the Guards, would automatically rally to the support of the Tsar-Emperor whenever necessary. It was also taken for granted that the Army would serve as a permanent, powerful, and wholly trustworthy backup to civilian law enforcement agencies. There was some criticism from minorities, but the Russian people, military and civilian alike, did not find it strange when Nicholas II kept his best troops in European Russia during the war with Japan, "more or less to sit on the heads of the Russian people," as one distinguished historian used to say. Officers and troops were also used, as was military law and courts-martial, to mop up after the 1905 Revolution. But, as noted before, the post-1905 army was a different army in many ways; and it changed even more drastically during the first two and a quarter years of World War I.

At the outbreak of the war the Army took full command over the combat zones and also over the areas to the rear of every zone. (Such areas were called military districts.) This did not work as neatly in practice as it had appeared in the Regulations of the General Staff. It saddled the military with responsibilities that they often were ill-prepared to handle. The capacity to cope with those responsibilities varied with the caliber of the officers and the civilian officials, as well as with local situations, but on the whole the situation created general confusion, conflicts among authorities, and many frictions. The military districts moved steadily eastward, first, as the fighting increased, and, later, as the Russian armies were forced to retreat. Most of western Russia, including Petrograd and its environs, was under partial military control by 1916.*

*Figure 1, below, shows the six "fronts" or armies, the Supreme Command, and the commanders of the land forces as of January/February, 1917.

THE FEBRUARY REVOLUTION OF 1917

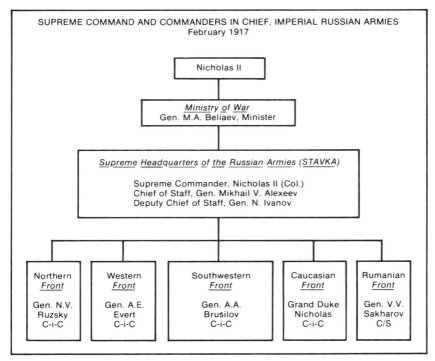

The Russian armies had numbered about 1.75 million officers and men on active duty when the war began. Full mobilization added 3.5 million reservists and militiamen to the armies' manpower. By the end of 1916, some 15 million men had been taken into the armed forces. This amounted to approximately 20 percent of the total male population; over 30 percent of the males of working age. Manpower losses were staggering, amounting to 3.8 million men by mid-1915 and escalating thereafter. The best estimates place the total losses up to March, 1917, at between nine and ten million. About 5.5 million new personnel were sent to the fronts during 1915-1916, and the High Command was estimating its manpower requirements for 1917 at 300,000 inputs per month. No such pool of available manpower remained. The Army, in other words, was physically different from what it had been at the start of the war. Unit names survived and so did some officers and men, but the new had submerged the old, and this had extremely significant consequences, first at STAVKA and Petrograd, and later at other places throughout the Empire.

The opening paragraphs of this paper at least hinted that it would apply some new perspectives to the February/March Revolution. The rationale is that analytical tools complementary to those long used by historians make it possible to approach the subject not as a single, seamless reality, but as a large series of what different persons and groups thought or felt to be the "real revolution." In other words, advancing knowledge of human behavior makes it clear that there was not just one real Russian Revolution. The real revolution to Maklakov was the abdication of Tsar Nicholas II, which Maklakov considered to have been "the last legal act" of the old regime. For Nicholas, however, the abdication was an anticlimax. His real revolution preceded the act of abdication.

When reports reaching STAVKA from Petrograd approached the hysterical, the Tsar asked Alexeev to consult with General Nikolai Ivanov and the Army Commanders. They unanimously advised Nicholas to seek reconciliation with those political leaders whom the Tsar regarded as only less dangerous to him and to Russia than were the Germans. He was psychologically unable to accept such a suggestion, and the rapidity of deterioration of civil order in Petrograd prevented any attempt to persuade him. It was entirely natural for the Tsar-Emperor to call upon his top military officers once again. He expected sound advice, which he could accept, concrete and practical help in meeting the crises. Here, in words quoted from the Tsar's diary, is how they responded when he told them that political leaders in Petrograd were demanding his abdication: "They said in substance that in order to save Russia and to maintain order among the troops this decision [i.e., abdication] was necessary. I consented."

This tired, lonely, worried man must have felt as if the solid earth had suddenly opened beneath his feet. If these generals,

THE FEBRUARY REVOLUTION OF 1917

bound to him by solemn oaths, long association, and all the rest refused to help him, where could he turn? He had no options left. This was the real revolution for Nicholas. No wonder that the final line in that diary entry read: "All around me there is only treason, cowardice and deceit."

As for the Grand Duke, and the other senior officers who associated themselves with this advice, the real revolution occurred when they individually and collectively decided that their obligation to Russia was more sacred than their obligation to a particular monarch. This change in traditional values came slowly to some, more quickly to others and, obviously, was not shared by all nor probably by a majority of officers. Certainly they did not all go as far as Alexeev did in his conversation with Maklakov. It needs to be added, that some of the officers at STAVKA and at Petrograd had concluded that Russia's problems had become too grave to be solved by the application of force; that any successes would be only temporary and would lead to greater violence. They were no longer ready to commit themselves or their troops for a man in whom they could not believe. The examples set by these commanding officers were not lost upon their subordinates, whose predispositions, pro or con, were catalyzed. This was also true, and more immediately apparent, in the nonfeasance of the military commander in Petrograd.

Early in 1916 a confidential report from the head of Moscow *Okhrana*, the secret security police, warned the Minister of Internal Affairs about growing signs of public resentment "against the person of the currently reigning emperor." The *Okhrana* made a more extensive report, summarizing detailed information from all over the Empire, in October, 1916. This report emphasized the general hardships, the failing morale, and the possibility that minor disorders in any of several major cities, including Petrograd, might trigger an intense and dangerous crisis. Peasants and villagers, as well as city people, it said, were increasingly outspoken in their criticisms of the government. They were not, the report emphasized, revolutionaries, but they were "political oppositionists." There is no documented linkage, but it seems possible—indeed, entirely plausible—that this *Okhrana* report led to the development of a new contingency plan for Petrograd. Such a plan was, in fact, prepared in detail and approved by the Emperor in November, 1916. It provided, in essence, that in case civil disorders in Petrograd should prove too much for the civilian police to handle, command responsibility would immediately pass to the Commander of the Military District. An order from the Tsar in January, 1917, considerably changed this district. Petrograd was removed from the jurisdiction of the Northern Front, then under the command of General N.V. Ruzsky, and established as a separate military district under General S.S. Khabalov.* Khabalov reported directly to the Minister of War. The head of the civilian police, General A. Balk, Prefect of Petrograd, reported directly to the Minister of

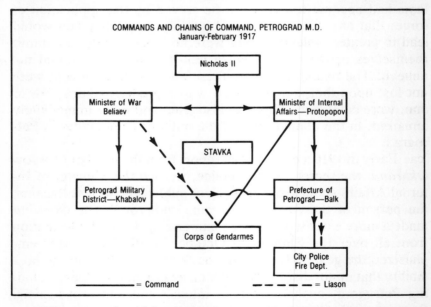

*The commands and the chains of command in Petrograd after this change was effected are diagrammed in Figure 2, above.

THE FEBRUARY REVOLUTION OF 1917

Internal Affairs. The Corps of Gendarmes, and its special branch, the *Okhrana*, were administratively responsible to the Minister of Internal Affairs and operationally responsible to the Minister of War. The heads of both of the latter agencies also reported directly to Balk. Since this new setup did not alter the contingency plan of November, a severe crisis would bring Balk and his people under the direct command of Khabalov and his forces. The facts belie Balk's later, courteous disclaimer of any friction between him and Khabalov. It was, however, a cumbersome system that might conceivably have worked had Khabalov and the Minister of War, General M.A. Beliaev, on the one side, and A.D. Protopopov, Minister of Internal Affairs and Balk's boss, on the other side, been of higher caliber.

This brings us to a consideration of the garrison troops in and around Petrograd, all of whom were Khabalov's responsibility, and the *ulitsy*, the slang phrase for the street people of Petrograd, who were Balk's responsibility until command was shifted shortly after noon on 24 February. The troops and the *ulitsy* were, for reasons that will be commented on later, not as disparate as might appear at first glance. It will suffice for the moment merely to observe that one without the other would have been less significant than it was though their alliances were ephemeral and informal. The garrison troops of General Khabalov were a motley crew, consisting of small permanent cadres, mostly reservists; convalescents and other limited duty types; some 1,500 officer candidates in training, mostly eighteen or nineteen years old; raw recruits, who ranged from nearly underage to nearly overage, assigned to various training units; and casuals in transit. (General Khabalov later reported to an investigating commission that his senior subordinate officer was suffering from chest spasms and that all his officers were ill, having been evacuated from combat zones for that reason.) Khabalov and Balk disagreed, then and later, about the strength of the troops commanded by the former, but there is no doubt that the garrison troops (depot troops in the official Russian terminology of that day) were a sorry lot, and this was not entirely their commanders' fault.

By 1916 the standards for acceptance into the armed forces had been greatly relaxed in order to provide the required number of bodies. The training period for recruits had been reduced in mid-1915 to a month or less, and the training had become seriously deficient in quality as well as in length. Any disciplined military organization must overlay the individual reality worlds of its recruits and volunteers with special values and assumptions peculiar to itself. There was neither the time nor the instructional talent to achieve this in the ever-swelling ranks of the Russian Army. Most of the new men were only superficially soldiers. They were trained and led, if they could be said to be either, by officers fit only for the most limited duty, by officers on temporary duty pending reassignment; by eight balls whom someone had managed to shunt aside; by the overaged and by the underexperienced.

The garrison troops, moreover, had a variety of legitimate grievances, large and small. They were overcrowded into obsolete, underheated, and ill-ventilated barracks. Their food was limited in quantity and poor in quality. (What became or, at least, what was called a mutiny in one outfit was sparked by the chant, "No more lentil soup!"—hardly a battlecry of revolution.) Garrison duty was dull and, more importantly, did not fully occupy duty hours. There were no provisions for the profitable or pleasant use of nonduty hours, and it is no wonder that idle soldiers wandered the streets, joined queues, followed and merged with crowds of *ulitsy*. What the soldiers saw and heard, as well as what they did, left them all the more susceptible to exploitation by agitators and others who were seeking to forward their own interests or the interests of some cause. Finally, there was a dreadful dearth of responsible leadership. Perhaps no commander could have done better than Khabalov, who subsequently could only describe the situation as having been "ghastly," but no leadership of any kind came from him. He not only did nothing; he didn't do this well. It would have required

THE FEBRUARY REVOLUTION OF 1917

incredibly farsighted, able, industrious, and lucky conspirators to have deliberately created such an unstable and potentially explosive situation. No such person or body of persons existed.

It is as impossible to provide accurate counts of the *ulitsy* as to provide such counts for the garrison troops. The vague generalization, "several hundred thousand garrison troops," was used above. This is a guess, and it is based on another guess, namely, that there were roughly 1.9 million garrison troops in the Russian Army in February, 1917, with perhaps a sixth of these stationed in the Petrograd area. The next set of figures gives an illusion of precision, but these data are also guesses and subject to accidental error as well as to deliberate inflation by the Bolshevik sources from which they were taken.

Isolated mutinies among units of the garrison troops first appeared on 25 February. These were not handled effectively by Khabalov or his people, and the infection spread like a prairie fire. The Bolsheviks place the number of rebellious or mutinous troops at 72,000 on the morning of the 28th, and at 170,000 by noon of the next day. (Khabalov claimed to have had less than 1,100 loyal troops under his command, but Balk challenged this count and asserted that Khabalov still had eight tanks, twenty-four field guns, and over 3,000 troops, not counting the officer candidates. Be that as it may, Khabalov behaved as if he had no troops; nor did his superior, General Beliaev, Minister of War, question Khabalov's nonfeasance.)

Petrograd had been changing into a new city in terms of the number of its people, the percentage of them who were gainfully employed, the percentage employed in industry, and the percentage of them who were literate. These and other advances had been slowly and unevenly but steadily providing new opportunities for Ivan Ivanovich (the Russian John Doe), with the usual resultant rise in ambitions and aspirations, as well as frustrations when further progress appeared stifled. The war brought floods of persons, in and out of uniform, into this industrial center, capital, and largest city in the empire. Most of the newcomers were peasants, taken into the armed forces or come to find better economic opportunities. They were in a very real sense, though not its usual meaning, displaced persons, both physically and psychologically. Housing was increasingly in short supply from the outset, and this condition worsened. As the war dragged on, other shortages—some temporary, others nearly permanent—developed. It was, in general, the new arrivals who fared least well. Wages rose, but prices rose faster. Queues became longer and longer at bakeries, markets, shoemakers, and the like. Shortages also affected industries. Mills and factories had to slow down or stop production when they ran out or ran dangerously short of fuel and raw materials. When this happened, the number of *ulitsy* was swelled by men out of work. Industrial conflicts including strikes and lockouts followed.

These hardships and tragedies were not in themselves except-for-which causes of the February/March Revolution. The critical mass, to borrow the phrase from high-energy physicists, was not reached until Ivan Ivanovich became convinced on two points. The first was perhaps brought into focus by a continuous stream of no-holds-barred attacks upon the government by the politically articulate opposition. The term covers, but was by no means confined to, radical extremists. Nor can it fairly be charged that the criticisms created the situation though they certainly called attention to it and effectively deprived the government of any benefit of doubt. Ivan's first conviction was that the regime was wantonly cheating him and his, thereby frustrating their legitimate aspirations; and his second conviction was that this might be reversed in either of two ways. One, the establishment could, if it only would, alter its course 180 degrees and provide higher wages, cheaper and more abundant food, better living and working conditions, and greater personal security for Ivan and his family. The majority of Ivans would have preferred this because their primary concerns were immediate and personal. A minority preferred the alternative and came to believe that it was attainable: namely, that the people could and should force the establishment to meet popular demands.

THE FEBRUARY REVOLUTION OF 1917

There was a true meeting of emotions, if not of minds, between the garrison troops and the *ulitsy*; more accurately, between minorities of both. They acted at first impulsively and individually and won some small successes, or so it seemed to them. Since they acted out of hope and not despair, each gain—even if it was only escaping any immediate, unpleasant consequences—fed their self-confidence and brightened the hope that greater gains might be theirs.* What brought the soldiers and *ulitsy* together? They had much in common because so many of them had come from similar backgrounds. The great majority of workers and soldiers were peasants or peasant descent. Some of the workers had been soldiers, and some of the latter had been workers in farms, fields, and factories. Some of the hooligans among the *ulitsy* differed little from the hooligans in uniform. When the Petrograd streets came alive with idled workers, with off-duty soldiers and sailors—some of whom were certainly AWOLs,—with people queuing for bread or milk or clothing, and with others of their kind, personal interchanges were myriad and electrifying.

Jammed together in huge crowds, sharing physical and emotional hungers, excited by happenings that caught them up and spiced their humdrum lives with adventure, troopers and *ulitsy* were highly suggestible, ready to follow any leader who gave tongue to their inarticulate, pent-up feelings. For the first several days the crowds went home at night and the streets and the city became quiet, but then the crowds became larger and more unruly, slipping now and again across the very narrow line that marks the vital difference between a crowd and a mob. Perhaps of greatest significance was a very special bond between barracks and street: a fear of consequences.

Military personnel absent from their units without leave and

*Various minorities and their leaders—anarchists, agrarian and Marxian socialists (Mensheviks and Bolsheviks) among others—tried with some success to manipulate the *ulitsy* and the garrison troops in order to advance the manipulator's selfish interests. The pictures become distorted, however, if one forgets that the interests of most persons were practical, short-range, and personal, and not revolutionary.

workers absent from their jobs in defiance of their bosses shared a terrifying prospect if the old order suddenly was returned. That meant punishment, in all probability very harsh punishment, which the absentees could hardly hope to escape. It was a frightening prospect for anyone who had been involved in any way, even by just his passive presence, in an illegal or punishable action. If he didn't think of it on his own there were many ready to remind him. Once this fear had taken hold, such persons were easily convinced that there could be no turning back for them. This was the final, indispensable component in the mix that bound garrison and street into what Dr. Bertram D. Wolfe has very accurately termed *stikhiia*—the uncontrollable, unpredictable, and perpetually fluctuating elemental force that brought the end of the Romanov dynasty and of Imperial Russia.

It is impossible to know what the real revolution was for the tens of thousands of the *ulitsy* and garrison troops because so few of them left any personal records. Supplementing these few by analogies drawn from more recent crowd/mob actions, one may risk a few conjectures. To some, the real revolution probably occurred when they committed with impunity acts which in more orderly days would have brought swift punishment. For others, the real revolution was a sudden burgeoning of hope as old restrictions were replaced, or appeared to be, by greater opportunities. For some of the military in Petrograd, as for General Balk, the Prefect, the real revolution came when the military command ordered officers and men to stack rifles and return to barracks. "Going through the corridors," [early in the morning of 28 February, wrote General Balk in his unpublished diary], "I was persuaded by puzzled officers returning to their quarters that they had received orders to lay down their arms and lead their units to the barracks. . . . General Khabalov was not to be seen." Finally, there must have been many to whom the revolution became real and meaningful when they persuaded themselves, or were persuaded by others, that they were now outlaws for whom there could be no turning back.

THE FEBRUARY REVOLUTION OF 1917

We have looked at the February/March Revolution not in its entirety; not as a chronological nor thematic narrative of incidents, triumphs, defeats, violence, and excitement. We have not replowed ground already fitted nor retrod paths, long worn smooth. We have made fresh approaches from new starting points by looking at the extremes of rank, power, and influence along the chain of military command from the Emperor to the nameless recruits of the Petrograd garrison. We have isolated and concentrated upon those personages and groups about whom it could be justly said, had they acted differently subsequent happenings would probably have been significantly different. We have tried to establish what the February/March Revolution really was to the Emperor Nicholas II, to his top commanders and a few of lesser place, and to the people of Petrograd and its streets whether in or out of uniform. We have sought explanations for their various behaviors to the extent possible within the confines of time and space available. We have stated our explanations not in terms of the operation of "blind forces," economic or other, but in the largely self-generated feelings and consequent actions of men.

═══════════ COMMENTS ═══════════

The more subtle sort of interplay between armies and polities comes to the foreground especially when a state chooses to employ the military arm as an instrument of diplomacy in more oblique fashion than through the usual coercion or deterrence. In his American Historical Review *article "Canada and the Siberian Intervention, 1918–1919" LXIV (July, 1959): 866–877, Gaddis Smith of Yale University showed how the attitudes of the Canadian military contingent in Siberia—not altogether unsympathetic to the revolutionaries, and certainly sympathetic to their American comrades in arms—undercut the purposes of the British who were employing them—purposes that ran counter to the American effort to keep the intervention strictly limited. The paper presented by Professor Smith to the New Dimensions course in 1973 displays some of the same kinds of crosscurrents dividing soldiers and civilians; not focusing on the Canadian intervention in this paper, but instead painting a wider canvas, Smith suggests still further difficulties of the employment of armed force in pursuit of complex diplomatic goals. Gaddis Smith has studied the intersection of the history of foreign policy with military force also in* Dean Acheson, Volume XVI of American Secretaries of State and Their Diplomacy *(New York: Cooper Square, 1972).*

Multilateral Intervention in Russia, 1918-1919: Three Levels of Complexity

GADDIS SMITH

PRESIDENT Woodrow Wilson said in 1918 that the problem of acquiring understanding and control of the situation in Russia reminded him of trying to capture under your thumb a drop of mercury on a sheet of glass. Even with the perspective provided by more than half a century, the situation remains complex and perplexing to an extraordinary degree. And yet this very complexity makes the subject of multilateral intervention in Russia at the end of the First World War a superb case study. The exercise of grappling, as students of history, with those events is useful for anyone whose professional responsibilities require that he analyze and make judgments concerning international affairs.

The scene can be quickly set. When the First World War broke out in 1914, Imperial Russia was allied with England and France (later joined by Italy) against Germany and Austria-Hungary. The Tsarist regime, barely able to govern before the war broke out, hoped the patriotic stimulus of combat would bring the nation together and save the monarchy. The Tsar and his advisers were wrong. The war produced the collapse of the inefficient Russian bureaucracy, the primitive transportation system, and the incompetently led army. In March, 1917, the Tsarist regime was replaced by a nominally democratic provisional government, soon headed by Alexander Kerensky. This event coincided with the entry of the United States into the war against Germany. Americans, ignorant of the depth of suffering and dislocation within Russian society, hailed the new government as "a fit partner in a league of honor." The Kerensky government tried to remain in the war—but to no avail. In November, 1917, the Bolsheviks, marching under the banner of "Peace, Bread, and Land," seized power and proceeded to take Russia out of the war.

The military implications of Russian withdrawal were, of course, catastrophic for the Western Allies. Germany, using its excellent internal lines of communication, was able to transfer enormous power from the Eastern to the Western Front. By the spring of 1918 a German victory was a real possibility. The French Army was still staggering from the impact of the mutinies of 1917. The British Army had been bled white by the ill-advised Flanders offensive of late 1917. The Italian Army, defeated at Caporetto, was a virtual nullity. German submarines had nearly cut the supply lines across the Atlantic. The United States was still training its own Army. No large American units had yet seen combat.

Desperate men in London and Paris looked for desperate solutions. Perhaps, they thought, some sort of intervention in Russia could topple the traitorous Bolshevik government and bring Russia back into the war. The British and French badgered their powerful associate, the United States, to approve of and join in the intervention.

President Wilson and his closest advisers believed in the beginning that intervention was absolute folly. They had a good instinctive grasp of the logistical difficulties of mounting military operations halfway around the globe in severe climatic conditions. With unhappy memories of the recent American experience in Mexico, they also were properly skeptical of the ability of military forces to bring about political change in a foreign country. They were opposed above all to the idea of using Japanese troops as a part of the intervening force in Siberia. Japan was an inactive ally in the war against Germany. Her participation had consisted of gobbling up some German islands and the German sphere of influence in Shantung Province of China. She had provided some minor naval assistance to the Allies and had enjoyed a very profitable war trade. Her large army was unused. But Japan and Russia were bitter rivals. President Wilson knew that the moment a Japanese soldier set foot on Russian soil, the effect would be to unite many factions of the Russian people against the intervention. Thus, for many months the United States government refused to participate in any scheme for intervention.

The above summary is misleading. It is too simple. Now let us turn to a closer look at the three levels of complexity that characterized the entire affair. From the point of view of the United States those three levels are:

1. *Who are your friends?* The United States was associated in the war with Britain, France, Italy, Japan, and China. All were interested in intervention—for very different reasons. Most of those reasons were incompatible with the larger purposes of the United States. Another participant was Czechoslovakia—not yet an independent nation in the physical sense, but rather a group of patriots seeking to win recognition for their right to independence in the peace settlement that would presumably follow Allied victory. Approximately 50,000 Czech soldiers were inside Russia in the spring of 1918. Originally they had been conscripts into the Austrian Army. They had changed sides in order to fight against their oppressor. Now they wanted to get out of Russia in order to participate in the fighting on the Western Front.

2. *Who is your enemy?* Was the Bolshevik regime in Russia an invention of the Germans, an instrument in a German drive for world conquest? Or was the regime the product of internal Russian developments, an expression of Russian self-determination? Or, most ominously of all, did the Bolsheviks represent an ideological disease that could sweep Western civilization from the face of the earth? Americans entertained all

three of these possibilities. Their inability to get accurate and consistent information about Russia and to reach even the beginning of a consensus about the nature of the Bolshevik regime represented a second level of complexity.

3. *What is your own policy?* Should American policy be defined in terms of the idealism of President Wilson's Fourteen Points? Should it be defined entirely in terms of maximum military advantage in the war against Germany? Or should policy be shaped in terms of postwar interests? All three of these positions were held by different American leaders at different times and by segments of the public.

Let us now look at each level of complexity in more detail. If the policies of all the nations with which the United States had to deal in the intervention question had been fixed, the complexity would have been deep enough. In fact, within each country there were divisions of opinion, conflict, and inconsistency. We know the most about British policy and opinion. Opinion on the British Left saw the Bolshevik regime as the product of internal conditions in Russia. They did not accept the idea that the Bolsheviks were tools of the Germans. They were also sympathetic to the Bolshevik analysis of imperialism and the causes of the war. They were even more sympathetic to the idealistic call of Woodrow Wilson for altruistic peace in the interests of all mankind. Therefore, they were deeply suspicious of proposals to topple the Bolshevik regime.

British Conservatives despised Bolshevism and wanted, above all, to see the regime destroyed. Many Conservatives also were deeply committed to the British position in India. They saw a powerful revolutionary Russia, possibly aligned with Germany, as a threat to India.

The British military, and especially the former military attaché to the Tsarist government, believed that intervention would be effective in rallying anti-Bolshevik forces. Like a mighty snowball rolling down a hill, they saw counter-revolutionary forces coalescing around a nucleus of outside intervention in order to bring Russia back into the war against Germany.

MULTILATERAL INTERVENTION IN RUSSIA

Prime Minister David Lloyd George was on a tightrope. He did not want to alienate left-wing British opinion, but he also wanted desperately to find an alternative to the interminable bloodletting on the Western Front. He had no deep understanding of Russia, but he was attracted to the military idea that intervention could turn the tide against Germany.

French opinion ran through a similar spectrum. The French, like the British and the Americans, had some observers in Russia who said that the Bolshevik government was not a threat and that cooperation rather than intervention was the proper course. But French middle-class opinion was violently anti-Bolshevik. The fact that the Bolshevik government had repudiated vast debts, money borrowed from the French people and government by the Tsarist regime, contributed to this hostility. The French military saw real opportunity in Russia, especially by using their protégés, the Czechs. Prime Minister Georges Clemenceau was a devoted anti-Bolshevik.

The Italians were not a major factor. Although they ultimately contributed a small contingent to intervention, their primary objective was to get the maximum Allied aid and military assistance in order to save their situation.

The Czechoslovak objective was their own independence. They realized that the 50,000 Czech soldiers in Russia were the best organized and potentially the most effective non-Bolshevik force. They wanted to behave in a way to maximize the support that the powerful Western Allies and the United States would give to their national aspirations.

The Japanese possessed, in their army, the greatest concentration of potential power for the intervention. They had economic and political objectives in Siberia. But at the same time the Japanese government was following a line of cooperation, especially with the United States. They sought a way of gaining their imperial objectives in Siberia without antagonizing the United States. The Japanese also wanted to secure the final removal of Russian influence from Northern Manchuria.

The Chinese government was in a chaotic state. The official government in Peking sought to use China's nominal participation in the war as a means of winning concessions at the peace conference. It sought also to protect its crumbling position in Manchuria against both Russians and Japanese.

There was yet another national participant in the intervention, one generally neglected by historians, Canada. The British, with almost no troops to contribute of their own, asked the Canadian government to furnish 4,000 men for intervention. Canadians agreed, and justified their participation with special and rather naive Canadian arguments. The Canadian Cabinet was persuaded that a vast economic opportunity in Siberia was at hand. Siberia and Canada were geographically similar. Canadians had unequalled experience in the development of natural resources in northern areas. Therefore, the presence of Canadian forces in Siberia might lead to Canadian opportunities to develop Siberian timber, mining, railroads, and fisheries. The Canadian troops who ultimately went to Siberia were actually supplied with little pamphlets entitled "Every Man a Trade Commissioner."

The problem of understanding whether the Bolshevik regime was an enemy of the United States and, if an enemy, of understanding its character was complicated by the general absence of Russian experts within the United States and by the conflicting reports received by the American government from Americans inside Russia. Before 1914 Americans had a general antipathy for the Tsarist regime. They understood vaguely that it was cruel and dictatorial. The existence of exile camps in Siberia was fairly well known. In addition, Tsarist persecution of Jews was a cause of bitterness (how little some things change!). The disappearance of the Tsarist regime was, as we have seen, greeted with general uninformed enthusiasm. During the short-lived Kerensky regime many Americans, in different capacities, went to Russia to report and to offer what aid they could to the Russian war effort. There were Red Cross officials, military attachés, propaganda experts, YMCA officials, financial and trade experts, and ordinary foreign service officers. Each person considered himself *the* expert on Russia and the man whose reports were shaping American policy. When the Bolshevik Revolution broke out each American developed his own viewpoint and set of recommendations.

Edgar Sisson, a newspaperman working for the American government propaganda agency (The Committee on Public Information), was persuaded that the Bolsheviks were invented by the Germans. Sisson paid a large sum of money for some bogus documents implicating the Germans in the creation of the Bolshevik regime. These documents were sent back to the United States and published authoritatively by the State Department.

Raymond Robins, an energetic Progressive and follower of Theodore Roosevelt, was in Russia ostensibly as a Red Cross official. He was persuaded that the Bolshevik regime was concerned primarily with Russian interests and was capable of cooperating with the United States. He devoted his energies to trying to bring this cooperation about.

David Francis, the American Ambassador, was an amiable and aging politican from Missouri whose reports had little importance. American consular officials in Russia, on the other hand, were better informed. They tended, however, to see things from local perspectives. The result was that they gave fairly heavy emphasis to the existence of anti-Bolshevik forces on a regional basis. Their reports helped form a generally anti-Bolshevik attitude in the Department of State.

American newspapermen reporting on Russia were no better informed than government officials. Their views, also, ran a gamut from sympathy to extreme hostility toward the Bolshevik regime. The weight of American editorial opinion, led by the *New York Times*, was overwhelmingly anti-Bolshevik. Congress, also, saw Bolshevism as a sinister force linked to radicalism and subversion in the United States. In 1918 there was actually a Senate committee that held hearings on the impact of Bolshevism on the German brewing industry in the United States.

Now let us turn to the third level of complexity—American policy. President Wilson's first important statement of policy following the Bolshevik Revolution was the sixth of the Fourteen Points of January 8, 1918. It is a long point and rather complicated, but it deserves quotation and analysis:

> The evacuation of all Russian territory and such a settlement of all questions affecting Russia as will secure the best and freest cooperation of the other nations of the world in obtaining for her an unhampered and unembarrassed opportunity for the independent determination of her own political development and national policy and assure her of a sincere welcome into a society of free nations under institutions of her own choosing; and, more than a welcome, assistance also of every kind that she may need and may herself desire. The treatment accorded Russia by her sister nations in the months to come will be the acid test of their good will, of their comprehension of her needs as distinguished from their own interests, and of their intelligent and unselfish sympathy.

At this stage, of course, the evacuation of Russian territory referred to territory occupied by the Germans. A plea for self-determination and a welcome of Russia "into the society of free nations under institutions of her own choosing" would seem to condemn intervention, especially if it sought to overthrow the Bolshevik regime. But did Bolshevism, in Wilson's mind, mean a set of institutions freely chosen by the Russian people? Or was Bolshevism forced on the Russian people? In Mexico, Wilson had tried to use military force in an effort to teach the Mexicans to accept democracy. Would he try the same in Russia? And could there be a welcome into "A society of free nations" of a regime which denied the very legitimacy of that society? Wilson was seeking a liberal reform of the existing international system. Lenin was seeking to abolish it through world revolution. In short, the sixth point confuses more than it clarifies. Wilsonian idealism provided few usable guidelines for American policy.

Was military arithmetic a better guide? The British and French claimed that a small military investment would reap enormous dividends in the war against Germany. Wilson and his military advisers were extremely skeptical. They suspected that the Allied military argument was simply a cover for selfish imperial purposes. All American military efforts were being directed toward the creation of a powerful, fully self-supporting, American Army for the Western Front. To divert any forces for a harebrained scheme in Russia seemed foolish. The Allied argument that intervention was also necessary in order to protect vast quantities of military supplies, previously shipped to Russia, from falling into German hands was not very persuasive either. If the Russian transportation system had failed to provide the means of carrying those supplies to the battlefield in previous years, there was little likelihood that the Germans could do any better.

And what of postwar American interests in Russia? Could intervention help advance them? The American business community was seized during the First World War with a touch of delusion about the Russian market. Business fantasies about Russia are strongly reminiscent of earlier fantasies about limitless profit to be found in trade with China. The business argument ran that before the First World War, Germany was the principal seller in the Russian market. Germany would be eliminated as a result of defeat. The war had also introduced the Russian people to the modern world and had created a tremendous consumer demand. What nation would be better equipped to move into this new market than the United States? The one remaining obstacle was the existence of a hostile ideology. President Wilson was partially sympathetic and partially opposed to this business argument. He wanted to see the United States develop markets in Russia in order to help the Russian people. On the other hand, he was deeply suspicious of the selfish interests of American businessmen who might exploit the suffering of the Russian people, to overcharge on their sales, and to underpay on their purchase of Russian raw materials.

Another long-standing American interest in Asia was to block Japanese expansion and to maintain the Open Door in China. The vision of large-scale Japanese intervention in Siberia was a great threat to this interest. Japan would win complete control of all of Manchuria, mobilize the resources of Manchuria and Siberia, and become the dominant power in East Asia. China would be converted into a satellite. The Open Door would be closed. Therefore, American policy ought to aim at keeping Japan out of Siberia or, if that was not possible, at keeping Japanese influence at a minimum.

The American decision to intervene in Siberia and the character of that intervention illustrates the interaction of these three levels of complexity. Throughout the spring of 1918 President Wilson and his closest advisers resisted all British and French arguments, except one—the alleged plight of the 50,000 Czechs in Russia. The Czechs were in the process of trying to extricate themselves from Russia via the trans-Siberian railway. They hoped to find shipping in Vladivostok and thence, ultimately, to make their way to France, where they could rejoin the forces fighting Germany. Moving along the railway in revolutionary Russia was a slow and risky process. The Bolshevik government was willing to see the Czechs leave, but it could not prevent local conflict along the railway line which impeded the Czechs. The British invented a picture, which they presented to Wilson, of the valiant Czechs besieged by half a million ferocious and well-armed former German prisoners. British military intelligence prepared some fantastic maps whereon they drew a thin line across 6,000 miles of the trans-Siberian railway representing the 50,000 Czechs. Above it they drew an enormous black line nearly an inch wide, stretching for 6,000 miles on the map, representing the German prisoners. Wilson was persuaded that the Czechs were in immediate danger of annihilation. He believed that intervention was necessary in order to clear the railway and evacuate the Czechs. But he did not want to appear to be acting in cooperation with the British and French, whose purposes he deeply distrusted.

Accordingly, the United States acted without consulting the British and French. The Americans proposed a very limited intervention in cooperation with the Japanese. The Americans and Japanese would each send 7,000 troops to Vladivostok for the limited purpose of securing the railway and saving the Czechs. The invitation to the Japanese was in the nature of proposing a firebreak against a massive forest fire. Wilson believed, naively, that controlled and limited participation by the Japanese would prevent the larger imperialistic incursion that he so much feared.

There is an economic side to Wilson's plans for intervention, which deserves more emphasis than historians have given it. As we have said, Wilson feared that uncontrolled private American business interests would exploit the needs of the Russian people. Therefore, he established a special American government trading agency. Its official title was the Russian Bureau of the War Trade Board, Inc. Incidentally, the secretary of this government corporation was a young War Trade Board lawyer with the assimilated military rank of major. His name was John Foster Dulles. The Russian Bureau wrote to hundreds of American companies and placed articles in the American commercial press indicating that it was ready to buy American products for sale in Russia. In turn it would buy Russian materials and sell them in the United States. All this would be done with an eye to Russian needs and on a nonprofit basis. The Russian Bureau set up a large office in Washington and sent agents to Siberia. In the end, it carried out almost no business in Russia. But its existence is indicative of the Wilsonian point of view.

The first American troops landed in Vladivostok in August, 1918. They were under the command of Major General William S. Graves, whose sole orders consisted of an idealistic memorandum written by President Wilson indicating that intervention was not against the Russian people and did not represent interference in the Russian people's helping themselves. The sentiments were good; the usefulness of the memorandum as a guide to action was minimal.

The American troops soon found themselves in company with 70,000, not 7,000, Japanese; about 3,000 British; about 2,000 French; a few Italians; about 4,000 Canadians; and some

Chinese; not to mention a kaleidoscopic set of Russian factions—some politically motivated, others hardly more than robber bands.

The Japanese were working with some of the less savory Russian factions in order to advance their own interests. The British and the French were organizing the Czechs, not for evacuation, but for a drive to the west toward Lake Baikal and the Urals. General Graves, sticking to the nonintervention spirit of the Wilson memorandum, sat tight in Vladivostok and refused to cooperate with the Allies. This obedience to orders resulted in his being condemned as a Bolshevik by some of the more vociferous Allied officers in Siberia.

On November 11, 1918, the armistice with Germany went into effect. The war was over. The original anti-German rationale for intervention was over. Now what should be done about intervention? Public opinion in the United States and Great Britain was not in a mood for a continued presence of troops abroad. President Wilson and British Prime Minister Lloyd George hoped that they could convene a conference of factions fighting against each other in Russia in order to settle the civil war and bring the Allied forces home. Prime Minister Clemenceau of France refused to allow any Bolsheviks to come to Paris for such a conference. An alternative site was selected—on the island of Prinkipo in the Dardanelles—but the anti-Bolshevik forces refused to attend.

Winston Churchill, now Secretary of State for War in the British Cabinet, recommended massive, all-out intervention by one million troops in Russia to destroy Bolshevism once and for all. Wilson and Lloyd George rejected this idea as utterly impractical.

A secret diplomatic approach to the Bolshevik regime was tried. William Bullitt, a young member of the American staff at the Paris Peace Conference, headed this mission in the spring of 1919. He went with the support of Wilson and Lloyd George and believed that he succeeded in drafting with Trotsky an acceptable basis for ending the civil war, providing aid and recognition for the Bolshevik government, and withdrawing the intervening troops. But by the time Bullitt got back to Paris, Lloyd George was under attack from right-wing press opinion in England for being soft on Communism. Wilson was preoccupied with other things. Neither Wilson nor Lloyd George, therefore, was willing to follow up the proposals Bullitt had brought back from Russia. Bullitt resigned his position in disgust and later condemned Wilson in public testimony before the Senate Foreign Relations Committee.

The intervening troops remained as long as they were locked in by the winter climate in Siberia and in North Russia—another theater of intervention about which we have not spoken. Gradually in 1919 and 1920 they were withdrawn—all save the Japanese. The last Japanese elements remained until 1922, when they were withdrawn in accord with agreements reached at the time of the Washington Conference on Far Eastern questions and Naval Arms Limitation.

Most Americans today know very little about the intervention of 1918-1919. A typical college student would probably be surprised to learn that American forces were ever on Russian soil. The typical Russian, on the other hand, knows a great deal about the intervention. Its importance is stressed and exaggerated in Russian history. A picture is painted of the Russian regime establishing itself in the face of powerful Allied military intervention. This, of course, is a distortion. The intervention was hostile, but it was also small-scale, extraordinarily confused, and ineffective. The United States and the Allies ought never to have intervened. But the outcome could have been considerably worse. Conflict among the intervening powers, the general fatigue of the First World War, and American restraint led to a fiasco. But it prevented a larger scale and bloodier intervention that would have increased the suffering and tragedy of the Russian Revolution, added to the cost of a tragic era for the United States and the European powers, and, on balance, created far more evil than good.

Bibliography

In addition to footnote citations for sources in Gaddis Smith, "Canada and the Siberian Intervention, 1918-1919," *American Historical Review,* LXIV (July, 1959), 866-877, other prominent works that might be consulted include:

Baerlein, Henry, *The March of the Seventy Thousand.* London: Leonard Parsons, 1926.

Bradley, John. *Allied Intervention in Russia.* New York: Basic Books, 1968.

Browder, Robert. *The Origins of Soviet-American Diplomacy.* Princeton: Princeton University Press, 1958.

Bunyan, James. *Intervention, Civil War, and Communism in Russia: Documents and Materials.* Baltimore: Johns Hopkins Press, 1936.

Churchill, Winston S. *The Aftermath: The World Crisis, 1918-1928.* New York: Scribner, 1929.

Coates, W.P. and Zelda K. *Armed Intervention in Russia, 1918-1922.* London: Victor Gollancz, 1935.

Francis, David R. *Russia from the American Embassy.* New York: Scribner, 1921.

Graves, William S. *America's Siberian Adventure, 1918-1920.* New York: Jonathan Cape and Harrison Smith, 1931.

Great Britain, Army. *The Evacuation of North Russia, 1919.* London: His Majesty's Stationery Office, 1920.

Great Britain, Navy. *A History of the White Sea Station, 1914-1919.* London: His Majesty's Stationery Office, 1921.

Halliday, Ernest Milton. *The Ignorant Armies.* New York: Harper, 1931.

Hoyt, Edwin P. *The Army Without a Country.* New York: Macmillan, 1967.

Ironside, W.E. *Archangel, 1918-1919.* London: Constable, 1953.

Kennan, George F. *Soviet-American Relations, 1917-1920.* 2 vols., Princeton: Princeton University Press, 1958-1959.

Luckett, Richard. *The White Generals: An Account of the White Movement and the Russian Civil War.* New York: Viking, 1971.

Manning, Clarence A. *The Siberian Fiasco.* New York: Library Publishers, 1952.

Morley, James William. *The Japanese Thrust into Siberia, 1918.* New York: Columbia University Press, 1957.

Noulens, Joseph. *Mon Ambassade en Russie Sovietique, 1917-1919.* 2 vols., Paris: Plon, 1933.

St. John, Jacqueline D. "John F. Stevens: American Assistance to Russian and Siberian Railroads, 1917-1922." Ph.D. dissertation, University of Oklahoma, 1969.

Schuman, Frederick L. *American Policy Toward Russia Since 1917: A Study of Diplomatic History, International Law, and Public Opinion.* New York: International Publishers, 1928.

Silverlight, John. *The Victor's Dilemma: Allied Intervention in the Russian Civil War.* New York: Weybright and Talley, 1970.

Stewart, George. *The White Armies of Russia: A Chronicle of Counter-revolution and Allied Intervention.* New York: Russell and Russell, 1970. Originally published 1933.

Unterberger, Betty Miller, ed. *American Intervention in the Russian Civil War.* Lexington, Mass.: Heath, 1969.

———. *America's Siberian Expedition, 1918-1920: A Study of National Policy.* Durham: Duke University Press, 1956.

U.S. Department of State. *Papers Relating to the Foreign Relations of the United States: 1919, Russia.* Washington: Government Printing Office, 1937.

Warth, Robert D. *The Allies and the Russian Revolution.* Durham: Duke University Press, 1954.

Welter, Gustave. *La Guerre civile en Russie, 1918-1920.* Paris: Payon, 1936.

White, John Albert. *The Siberian Intervention.* Princeton: Princeton University Press, 1950.

====== COMMENTS ======

Military government of occupied territories is one of the most overt forms of political activity by the military. It is so overt that states committed in principle to thoroughgoing civilian control of the military only reluctantly admit that they must engage in it, as Earl F. Ziemke showed in discussing the United States and the Soviet Union in his 1973 presentation to the New Dimensions course. In the United States, military government has special unhappy overtones, furthermore, beyond those that Professor Ziemke specifies; the United States Army's first extensive experience as an army of occupation, after a brief and successful venture in Mexico, was in the former Confederate States after the Civil War. There the somewhat paradoxical combination of having to occupy part of the United States itself and the special bitterness felt and displayed by the conquered toward the conquerors created a legacy of distaste for occupation duties that, however vaguely remembered, has never been entirely dissipated in the American Army. Occupying the South during Reconstruction also inevitably entangled the Army in the acerbic politics of the period, which underlines why soldiers and civilians alike tend to approach military government with so much chariness: the directives guiding the army of occupation can never be specific enough to save the army from the continual exercise of highly political judgments.

Professor Ziemke of the University of Georgia has contributed to the United States Army's histories of World War II, **Stalingrad to Berlin: The German Defeat in the East** *(Washington: Office of the Chief of Military History, 1968).*

Military Government: Two Approaches, Russian and American

EARL F. ZIEMKE

THE PURPOSE of this essay will be to undertake the improbable feat of comparing something that does not exist with something that it is strongly believed should not exist, namely, Soviet and United States military government. The term "military government" as such does not exist in the Soviet military lexicon. They have used "military administration," but even it has been applied only once, perhaps, in an actual situation, and that in Germany after World War II when the Western Powers were setting up so-called "military governments" in their occupation zones. Otherwise, however, there is no evidence that the Soviet Army assumes government or administration of occupied territory to be so much as one of its peripheral functions. In the enormous Soviet literature recounting the military achievements of World War II, the occupation of Germany is barely mentioned, and nothing at all is said about the occupation of the eastern European countries, Korea, or China.

On the other hand, the United States Army Judge Advocate General, in the summer of 1940, actually published a field manual, FM 27-5, with the title *Military Government*. But the words fell rather too harshly on almost everybody's ears and soon gave way to the more indefinite "civil affairs," which had been used during the post-World War I occupation of the German Rhineland. "Military government" raised visions of empires and proconsuls, of Antony and Cleopatra, of Genghis Khan, and even of Attila the Hun. Once the United States was in the war, though, some taste of these seemed appropriate for the Germans and Japanese and Italians; so civil affairs became military government in occupied enemy territory. By the 1950s, although military government had been reasonably successful in both Germany and Japan and had not perpetrated any demonstrable political, social, or moral outrages on the occupied countries, it again fell into disrepute. The Military Government School became the Civil Affairs School and the field manuals became civil affairs manuals. FM 41-10, *Civil Affairs Operations,* published in 1969, defines civil affairs as: "Matters concerning the relationship between military forces located in a country or area and the civil authorities and people of that country or area usually involving performance by the military forces of certain functions or the exercise of certain authority normally the responsibility of local government." The term "military government," by my count, occurs only once in the field manual, that is, when it is described as: "the form of administration by which an occupying power exercises executive, legislative, and judicial authority over occupied territory." The inference to be drawn, and probably intended, is that military government is now a subsidiary and exceptional aspect of civil affairs.

Probably, it seems by now that this is all just quibbling over words. Civil affairs is, or readily can become, military government. Military administration is military government. The United States Army has conducted military government in every major war it has fought since the war with Mexico in the 1840s and would most likely do so again if circumstances required it. The Soviet Army did the same in Germany and North Korea in the 1940s, may have later in Poland and Hungary, and also would do so in the future. In fact, however, this apparent quibbling is crucially significant for military government, or civil affairs, or military administration, and because they are potentially the most important of them, for the noncombat roles of military forces in general. It is significant because it denotes the existence of a frontier, a somewhat turbulent and perilous one at that, along which the military operates with substantially less certainty than it does in its recognized sphere. The conditions

MILITARY GOVERNMENT: TWO APPROACHES

that determine that frontier are different for the United States and Soviet Armies, but important for both.

The United States Army, as noted earlier, has conducted military government in all of its major wars since the Mexican War, provided we can agree that Vietnam and Korea were not major wars. In all except World War II, it was an afterthought, an improvisation that both the civilian and military authorities professed not to want. Colonel Irwin L. Hunt, Third Army civil affairs officer in the Rhineland, pointed this out to the War Department after World War I. In his final report on the Rhineland occupation, he recommended that since some form of military government seemed to be an inevitable concomitant of wars, the Army ought to prepare officers in peacetime for such missions. But the 1920s were a time when the Army's occupation of foreign territory seemed a good deal more remote than it does even now, and Hunt's recommendation would probably have vanished outright had not War College committees occasionally resurrected his report. It seems to have been the G-1 committees that were the most interested, because of the personnel implications.

Entirely unwittingly, Colonel Hunt and the War College committees had laid a kind of a time bomb in the form of a potentially explosive question, namely, did the government and the Army want to accept and prepare for as an established function something they had in the past always regarded as an emergency expedient? The 1934-1935 and the 1939-1940 War College classes helped by writing a draft military government manual and a manuscript on the administration of occupied territory. In the fall of 1939, Major General Allen W. Gullion, the Judge Advocate General, turned down a recommendation, also from the War College, that the Army publish a military government manual. The reason he gave was that the recently published FM 27-10, *Rules of Land Warfare*, gave all the guidance the Army needed. Early the next year, however, with further urging from G-3 and G-1, which supplied the War College materials, Gullion's office wrote FM 27-5, *Military Government*.

FM 27-5 established a requirement for military government

training, but nobody wanted it. G-1, assigned the staff supervision, was strictly a staff agency with no capacity or facilities for training; and none of the other elements of the General Staff could see any use in diverting men and effort from the Army buildup to prepare for what then appeared to be a postwar mission. But in the meantime, governments were disappearing in Europe, crushed or driven into exile by the Germans, and it was becoming clear that the current war would not be fought as World War I was in France, on the territory of an ally fully capable of handling its own civil affairs. Consequently, in January, 1942, on G-1's urging, the Chief of Staff, General George C. Marshall, overrode lingering nonconcurrences from other staff agencies and authorized General Gullion, who had in the meantime become the Provost Marshal General, to establish a military government school.

When Gullion opened the Military Government School in May, 1942 on the campus of the University of Virginia in Charlottesville, the bomb went off. The newspapers dubbed it a "school for *Gauleiters*." The civilian departments demanded shares in any future occupations, and the Army commands demonstrated their lack of interest by attempting to unload their misfits on the Military Government School. Worst of all President Franklin D. Roosevelt pronounced the governing of civilians anywhere to be a civilian task and let it be known that he did not think the military capable of governing under any circumstances. He proposed, in fact, to install civilian administrators in all occupied areas and make them responsible to the State Department or some other civilian agency. For the Army this was disaster. It could mean that every theater commander would have standing beside him a coequal civilian in a separate chain of command, practically what Clausewitz had said was the worst possible situation—two commanders on the same battlefield.

The ensuing conflict was, to my knowledge, never actually resolved. In North Africa, the President's view seemed to prevail as several civilian agencies attempted to conduct civil affairs

MILITARY GOVERNMENT: TWO APPROACHES

while the graduates of the Military Government School sat at home or in England without assignments. North Africa did demonstrate one thing however, namely, that only the Army had the men and equipment and organization to accomplish on short notice all that needed to be done to administer occupied areas overseas. As a consequence, President Roosevelt in November, 1943 gave the War Department as much of a charter to conduct military government in World War II as it ever had, when he informed Secretary of War Henry L. Stimson that the Army would have to assume "the initial burden . . . until civilian agencies [were] ready to carry out the long range program."

After Harry S. Truman succeeded to the Presidency in April, 1945, he asserted at one of his first briefings on occupation plans that civil government was "no job for soldiers," and the War Department then began preparations to turn over its civil responsibilities in Germany (Japan was not yet occupied) to the State Department by the end of the year. Having defended the single-commander principle through the war and finding civil government a public relations liability and a drain on its shrinking manpower, the Army was ready, even eager, to give up the job. The State Department had discovered in the meantime, however, as the civilian agencies had in 1942 and 1943, that it did not have the men or resources to govern twenty million Germans, much less a hundred million Japanese and some millions more of Austrians and Koreans. In January, 1946, Truman ordered the War Department to retain executive responsibility and the State Department to assume the policy of planning for occupied territories; and that was where the matter stood until the end of the post-World War II occupations. Since then, the doctrine appears to have reverted to the pre-1945 position. FM 41-10, of October, 1969, asserts military predominance in occupied territory during the combat phase but assumes a subsequent and early transfer to civilian agencies.

A good deal less, of course, is known about the historical development of Soviet thinking on military government. One reason is that officially no such thing could exist. Nothing could be farther from the Soviet mind than to prepare to occupy and rule other peoples' territory. Another, and also valid, reason is that there is probably less to be known. The Red Army operated as an occupation force in parts of Russia during the Civil War, and reputedly carried out similar missions in some places within the Soviet Union during the 1930s. In 1939, the Soviet Union occupied a good third of Poland; and in 1940, it took over the Baltic States, Bessarabia, and parts of Finland. But in all these instances, except for one or two small areas in Finland, the Soviet Union maintained from the outset that it was merely repossessing its own territory. Later in the war, when the Americans and British were planning the combined occupation of Germany, they several times invited the Russians to send a staff to join them in London. None ever came. The Soviet government, probably with intended irony, said it was too busy winning the war to spare men for postwar tasks. That the Russians had actually not prepared for military government during the war, at least not in the same sense that the Americans and British had, seemed to be confirmed in 1945. When the Anglo-American forces drove across Germany they had with them a number of special teams that were to have made contact with elements of Soviet military government. They never found any. The first such contact did not occur until several days after the German surrender, when the SHAEF control party that had taken over the remnants of the Nazi government under Admiral Karl Doenitz was jointed at Flensburg, Doenitz's headquarters, by a counterpart Soviet control party. The SHAEF people had been trained and ready for their mission since early in the year; the Russians appeared to regard a formally organized military government as a new and interesting idea and asked numerous questions about how the Western Allies proposed to operate it. Later, in the first week of June, when General Dwight D. Eisenhower and Field Marshal Bernard L. Montgomery met Marshal Georgi K. Zhukov in Berlin, they concluded that he did not yet have a top-level staff for military government in Germany of the kind they had been organizing and training for more than a year.

MILITARY GOVERNMENT: TWO APPROACHES

That the Soviet Army apparently did not have a corps of military government specialists, however, does not mean that the Soviet Government was not prepared to achieve its objectives in the occupation as it saw them. A day or two after Berlin fell, Zhukov appointed Colonel General Nikolai N. Berzarin, an army commander, city commandant of Berlin. Similar appointments were made in the other east German cities as they were occupied and, apparently, also in the rural districts after the surrender. The commandants quickly appointed officers from their own forces to take charge at the lower levels. The latter had nowhere near the technical expertise of American military government officers, but their authority vis-à-vis the Germans was so complete, and often arbitrary, that they did not need to be experts to get what they wanted done. On the day after the surrender, Anastas Mikoyan, the Politburo member who had supervised economic relations with Germany during the period of the Hitler-Stalin Pact, appeared in Berlin at the head of a large staff of economic and industrial specialists. Before he left Germany, some weeks later, he established his men throughout the Soviet Zone. They apparently had two tasks: one was to help the Army organize the economy for its own benefit and for the Germans; the other was to supervise dismantling of German plants and equipment to be shipped to the Soviet Union as reparations.

In one respect the Russians had a particular advantage over the Western Allies. They had a significant number of native German Communists who were fully and reliably committed to the Soviet interest. This was true also for the other areas the Soviet forces occupied, eastern Europe and North Korea. In Poland, for instance, they were able to install a complete Communist national government even before they had occupied any significant part of the Polish territory. Some of the people involved were Communists who had survived the war at home, underground or in jails and concentration camps. Fewer in total numbers but more reliable were those who had lived in the Soviet Union. Some of them had political careers dating back to before the Russian Revolution, the German Walter Ulbricht, for example, or the Hungarians, Imre Nagy and Matyas Rakosi, or the Korean, Kim Il Sung. Others were younger men and women, the sons and daughters of foreign Communists, who had in some instances lived almost their whole lives in the Soviet Union. All of them had been groomed for years for roles they were to play eventually, not in the Soviet Union but in their native countries. Some had acquired experience and seasoning in the Spanish Civil War or, as Ulbricht did, in the Soviet Army during World War II. The younger men and women were trained in Comintern schools and in institutes for their nationalities. They did not, like the students in American military government schools, learn the techniques of public administration; but they were experts on Communism as the Soviet Government wanted it understood, and they were flawlessly groomed in languages and the other requirements for them to fit into the life of their native countries. They were also as politically reliable as the Soviet methods of selection, indoctrination, and surveillance could possibly make them. The so-called National Committee for a Free Germany, formed after the Comintern was disbanded in 1943, which appealed also to German prisoners of war, particularly officers, incorporated the German Communists into an ostensibly German national movement.

Democracy, certainly, also had its sympathizers in the enemy and occupied countries, but they were a less homogeneous group. The United States also harbored large numbers of emigrés, particularly from Germany, but it insisted on using them strictly as American citizens, which was what most of them wanted. At the same time, the United States authorities were unwilling to use nonnative Americans in policy—or decision-making capacities; so most were employed primarily as linguists, that is, as interpreters or as information specialists. The Russians, on the other hand, were able very early, in Germany as early as the summer of 1945, to give the appearance of having turned over government operations to Germans—their own Germans, of course. The same was true elsewhere. In North Korea, the Soviet commander, Colonel General Terenti Shtykov, was already styled "ambassador" early in 1946. The

Russians, naturally, did not relinquish any actual power. Through the occupation forces and probably even more through the secret police, the NKVD, they kept their hand in at all levels.

It can then be said, on the basis of the comparison thus far, that although both the United States and the Soviet Union have in the past and may again find it necessary, military government is not a function that fits easily into the American or Soviet concepts of military roles and missions. From the American point of view, the chief sources of uneasiness are an ingrained assumption that the military mind is not and cannot be attuned to civilian affairs and a dark suspicion that any exercise of governmental authority by the military upsets the constitutional subordination of the military to the civilian authorities. Alongside these can be put also a strong military reluctance to be bothered any more than is absolutely necessary with civilian concerns. Basically, these are emotional reactions, the product of the way Americans look at themselves. To the extent that they represent not only national opinions and prejudices but generalized differences in military and civilian outlooks, they can probably also be applied to the Soviet Union. However, in the case of the Soviet Union some other significant considerations are involved as well. For one, the totalitarian state, although it relies more on military power than a democracy does, regards its army from a practical point of view as a relatively ineffective instrument for controlling its own or other people and prefers to rely on police and political organizations. In other words, where in the American view, military government implies a ruthless, mindless efficiency, in the Soviet view it is not efficient enough. Secondly, whereas the American tendency is to regard the military as uninstructed and unteachable in matters civilian, the Soviet inclination might well be to suspect that the military would learn only too well. One is reminded of the early nineteenth century when the King of Prussia long and strenuously opposed the idea of a citizen army. If the citizens were taught to march and use arms, he argued, how could he be certain they would not use those skills against him? If the Soviet Government allowed its army to develop the capacity for government, how could it be certain they would not some day use the capacity at home?

Having examined the American and Soviet conceptions of the military role in governing occupied territory, we can now attempt some comparisons in terms of purposes and performance. For the United States Army the problem has been, and is, to tailor military government to a pattern that will cover the military necessities and stay within the bounds of the military-civilian relationship. After World War I, Colonel Hunt earnestly recommended that the first purpose of military government in the future be "to make friends of former enemies." More than a few of his World War II successors in military government would have agreed with him, and the first edition of FM 27-5 (1940) included his recommendation among its stated objectives. Parenthetically it might be added that the Russians have, since World War II, touted as one of their greatest successes their army's ability to conquer the hearts of people everywhere it went.

But the emotional climate of World War II was—to some extent unfortunately—not conducive to maintaining Hunt's broad altruism. This, by the way, touches on one of the persistent problems of United States military government planning: Military government has to set its course in the midst of hostilities and do much of its work in a different atmosphere after the fighting has ended. In any case, the Army needed a more specific definition of purpose than Hunt's. The first such, adopted in 1940, was predominantly legal. Under international law and the rules of ground warfare, the Army had certain obligations to civilians in occupied areas, among them to maintain law and order and prevent damage to civilian persons and property. Carrying out those obligations would be the military government's responsibility. During the war, the occupation of friendly, so-called liberated, areas added requirements for relief, health and welfare services, and reconstruction. To keep them

MILITARY GOVERNMENT: TWO APPROACHES

from ballooning unmanageably while also assuring that they would not be ignored, the Army then made military government a command function, at the same time limiting it to the legal obligations and such other activities as would keep the civilian populations from becoming either a burden on or a threat to the tactical mission. The assumption was that comprehensive welfare or other programs would be the business of other agencies later.

Although, undoubtedly, many of the military government personnel saw themselves in a political role as apostles of democracy, this was not stated officially as one of their functions. Non-enemy countries were assumed to be democratic. In enemy territory, the task was seen as being to remove the antidemocratic and militaristic elements in the belief that democracy would then emerge. Until the Cold War began in earnest in 1946, the approach was to let the Germans, possibly less so the Japanese, find their own way to democracy and to assist primarily by preventing their backsliding into their former ways. Even at the height of the Cold War in 1948, President Truman maintained that the United States' aim was not to impose its political system on anybody, only to make certain that the right of peoples to choose their form of government was not foreclosed.

The purposes of United States civil affairs/military government as presently defined in FM 41-10 clearly derive from the World War II experience. The first is to support the conduct of tactical operations. The second to fulfill the commander's legal obligations to civilians. The third, and last, is to act as the military agency for the attainment of United States national objectives during and beyond the period of conflict. The last could encompass a range of possibilities but in fact, tends mainly to underscore the restrictive nature of the first two purposes. FM 41-10 further envisions a predominance of United States civilian agencies in Cold War and stability operations and an early shift to civilian control in combat situations.

It has been plausibly argued that the United States entry into World War I in 1917 and the Communist Revolution in

Russia in the same year launched both nations on campaigns in which each has tried to remake the world in its own political image. Therefore, it is also argued, everything either power does in the world is inevitably a self-serving political act. These arguments, however, overlook an important distinction. The World War I slogan, "to make the world safe for democracy," was a good deal more profound than it is given credit for having been. As it implies, the United States fought World War I—and World War II and entered the Cold War—not to impose a system but to keep an open world, admittedly one in which it hoped to see its conception of democracy take root and thrive. On the other hand, Stalin told Milovan Djilas in April,1945, "This war is not as in the past; whoever occupies a territory imposes on it his own social system. Everyone imposes his own social system as far as his army can reach. It cannot be otherwise." Even if Stalin had not been who he was, even if he were, for instance, Mr. Leonid Brezhnev, he could not have thought otherwise. Soviet Communism does not assume an open world but rather an eventual world government. This is where the difference in the purposes of U.S. and Soviet military government lies.

United States military government can accept international law and the customs of war as its obligatory guidelines. Soviet military government can only do so to the extent that those things coincide with Soviet doctrine and interests. It would be naive to pretend that United States military government has not and will not on occasion disregard the rules in what appears to be the national interest. The difference is that Soviet military government probably will not enforce the rules unless they are specifically in the Soviet interest. During hostilities, the United States theater commander is as free, within the scope of his mission, to direct military government as he is to conduct tactical operations. The Soviet Commander is, or at least has been in the past, under political surveillance for all his acts, and the record of World War II strongly suggests that in civil affairs, his authority is mostly *pro forma*. He and his forces merely provide the muscle and the visible presence behind the political and police authorities. Since it is hard to imagine anything more dangerous than for a Soviet commander to assume the power of political decision making, he is probably happy to have it that way.

In effect, what happens under the Soviet system is that the peoples of occupied territories are in fact not subjected to military government in the sense that Americans conceive it, namely, as a temporary condition in which the military forces assume civil authority in place of a government that cannot function or has ceased to exist. Where Soviet forces go, a political authority, the Communist Party, always exists, and it is by Soviet definition universal. Consequently, the people in occupied territory cease almost immediately to be objects of military concern and become the political subjects of communism. As Hanna Arendt pointed out in her work on totalitarianism, they are really regarded less as foreign enemies than as domestic rebels.

The differences in purpose also have a direct bearing on the performance of United States and Soviet forces in administration of occupied territory. The American tendency has been, and probably still is, to emphasize tactical support and the legal obligations—during hostilities even to the exclusion of everything else. The World War II field manual and directives made United States military government not merely nonpartisan but completely nonpolitical. The field manual, FM 27-5, prescribed an absolute prohibition on all political activity. This astringent approach had some advantages. It kept military government from becoming entangled with or falling captive to political factions—which is what it was, no doubt, primarily intended to do. It also tended to reduce partisan political activity which could have interfered with the tactical mission or impaired law and order. Further, it presumably left military government free to seek out the best available technical and professional help in occupied areas. On the other hand, it meant that Americans had to do much of the work and take all of the blame for what went wrong. It also gave rise to an image of Americans as not really knowing what they wanted. When there were no acknowledged in-groups, everybody was in the out-group. It tended, too, to slow revival of political activity after the prohibition was lifted. People became hesitant or reluctant to do for themselves what had been done for them, and military government officers were often unwilling to relinquish power and jobs. Lastly, to avoid political entanglements was never as simple a matter as it had appeared on paper in the first place. Native citizens still had to be found and employed, and somebody's word had to be taken as to their qualifications. Many a military government team learned eventually that the local Catholic priest or Protestant minister was not necessarily a more dispassionate collaborator than some overt politician.

Where comparisons could be made—primarily in Germany after May, 1945—the Soviet performance appeared initially to be vastly more flexible and effective than the American. The Russians had a cadre of German administrators for Berlin on the day the city surrendered, May 2, 1945. By early summer of 1945, the whole government of the Soviet Zone was nominally in German hands. The Americans did not begin turning back responsibility to the Germans until the end of the year. In June, 1945, the Russians licensed political parties in their zone. The Americans took another four months just making up their minds about the licensing procedure. Soviet-controlled Radio Berlin was broadcasting operettas and dance music for months while United States military government mulled over what kind of entertainment, if any, the Germans should have. While United States military government courts were clogged with minor cases, the Russians appeared to have no law and order problem. They had simply appointed German block leaders and made them answerable for their neighbors' behavior. In serious cases, their local commandants could pass sentences up to the death penalty on their own authority. In the summer and fall of 1945, American newspapers and news magazines, *Time* particularly, rated the Soviet performance in Germany well above that of United States military government. Popular opinion polls in the United States Zone also showed the Germans as being more impressed by the Soviet performance than by the American. Few of the Germans, however, were sufficiently impressed to want to move to the Soviet Zone, and one of the rumors that periodically threw the Germans into a near panic was that the Americans were going to pull out and leave their zone to the Russians.

A generation later, we can now also make some broader judgments. The chief weakness of United States military government has been that it has lacked, certainly in World War II, a strategic, that is, a political objective. The immediate fault in that regard lies with the reluctance of the United States political authorities to assign to the military a more than temporary and limited responsibility for civil affairs and the reluctance of the military to assume any more than the minimum of such responsibility. More fundamental and possibly more important has been the limited view of strategy in general. In World War II the United States practically closed shop on strategic planning after the Tehran Conference of December, 1943, where the decisions on the cross-Channel attack and, more tentatively, the Soviet entry into the war against Japan had been made. The implied assumption that strategy ends when the shooting stops effectively left United States military government with no other missions than those relating to the war itself or deriving from the war. The latest doctrine on military government, as stated earlier, does now take into account also national objectives beyond the period of conflict. This could alter the whole approach to military government, but one doubts that it has. The predominant emphasis is still on tactical support and the legal obligations.

Soviet military government has always had a strategic objective, namely, to extend the Communist system as far, as Stalin said, as the armies reach. The Soviet approach, however, also has a weakness, specifically, the inability to define Communism in other terms than narrow Soviet national interest and the not very attractive Soviet system. As a result people are still running away from East Germany at a rate of hundreds a month, and the Soviet Army has to intervene in force periodically to rescue the situation in the eastern European countries. It could also be argued that the outright self-serving nature of Soviet policy since World War II has gone a long way toward compensating for the initial unpreparedness of the United States, certainly in western Europe, maybe even in Japan. Whether or not the Soviet Army conducts military government, the world has learned and, hopefully, will not soon forget that it is an agency of Soviet imperialism.

Bibliography

Arendt, Hanna. *The Origins of Totalitarianism*. New York: Knopf, 1959.

Clay, Lucius D. *Decision in Germany*. Garden City: Doubleday, 1950.

Davidson, Eugene. *The Death and Life of Germany*. New York: Knopf, 1959.

Friedrich, Carl J., ed. *American Experiences in Military Government*. New York: Rinehart, 1948.

Friedmann, W. *The Allied Military Government of Germany*. London: Stevens and Sons, 1947.

Gardner, Michel A. *History of the Soviet Army*. New York: Praeger, 1966.

Garthoff, Raymond L. *Soviet Military Policy*. New York: Praeger, 1966.

Hart, B.H. Liddell, ed. *The Red Army*. New York: Harcourt, Brace, 1956.

Leonhard, Wolfgang. *Child of the Revolution*. Chicago: H. Regnery, 1958.

Nettl, J.P. *The Eastern Zone and Soviet Policy in Germany, 1945-50*. London: Oxford University Press, 1951.

U.S., Department of the Army. FM 41-10, *Civil Affairs Operations*. Washington: Government Printing Office, 1969.

U.S., Department of State. *North Korea: A Case Study in the Techniques of Takeover*. Washington: Government Printing Office, 1961.

U.S., War Department. FM 27-5, *Military Government*. Washington: Government Printing Office, 1940.

Zhukov, G.K. *The Memoirs of Marshal Zhukov*. New York: Delacorte Press, 1971.

VI

Unpopular and Unconventional Wars

The War of 1812
and the Mexican War
HARRY L. COLES
page 311

A Case Study
in Counterinsurgency:
Kitchener and the Boers
THOMAS E. GRIESS
page 327

Revolts Against the Crown:
The British Response
to Imperial Insurgency
J. BOWYER BELL
page 358

COMMENTS

The first three years of the New Dimensions course coincided with the United States' extricating itself from its military involvement in Indochina. Americans often assumed in that unhappy time that the unhappiness of the time was unique —that no war in American history could approach the Indochina War in unpopularity and in volume of dissent from the war. Many of the Army War College students themselves assumed that when they had been fighting men in Indochina, they had been uniquely denied the full measure of support of the American public.

Those students and most of the rest of the country were, of course, measuring their Indochina war experiences against what they remembered or had been told of the Second World War, when there had indeed been almost unanimous support of the war. But it was the Second World War, not the Indochina War, that was nearly unique in the history of American wars. Rarely have the nation's enemies in war seemed so patently worth fighting, rarely has a war seemed so plainly necessary in defense of the American way of life, as in the Second World War—especially because of the Japanese attack on Pearl Harbor. Every other American war has been to some marked degree an unpopular war. In fact, even the apparent straightforwardness and simplicity of a war's aims, such as the United States enjoyed in combating Hitler's Nazism and Japanese militarism in World War II, are evidently no assurance of consistent support for a war—unless the war, once initial disarray is overcome, exhibits consistent progress toward victory. It was the absence of serious setbacks, the consistency of American progress, after Pearl Harbor and the initial defeats associated with it that was probably at least as responsible as the evident malevolence of the enemies for the popularity of World War II. The historical record suggests that the unanimity of popular support for an American war is directly proportionate to the war's shortness and to the consistency of the success of American arms.

===== COMMENTS =====

Any war tends to become unpopular as it becomes prolonged and as military success is interrupted. Though the vividness of recent memory suggests that the Indochina War was phenomenally unpopular, in fact the evidence of public opinion polls indicates that the Korean War was even less popular—until, the crucial fact, the Indochina War had gone on longer than the Korean War and exacted more American casualties. Since almost all wars, in the post-Napoleonic age of indecisive warfare, become both prolonged and lacking in consistent military success, almost all wars have been unpopular wars.

In 1971, during the first year of the New Dimensions course while the Indochina War was still being fought, Harry L. Coles of Ohio State University reminded the Army War College students of the unpopularity of some of the past American wars. Professor Coles' work has ranged over practically the whole span of American military history; he has recently been studying the nuclear strategists, but he is also the author of a one-volume history of one of the unpopular wars he discusses here, **The War of 1812** *(Chicago: University of Chicago Press, 1965). In 1974–1975 he was the first holder of the visiting professorship in military history at Fort Leavenworth.*

The War of 1812 and the Mexican War

HARRY L. COLES

THE TOPIC, "Unpopular Wars and Military Operations," is not self-explanatory. Herein will be discussed the question of unpopularity or dissent in American military history and how lack of popular support actually affected the conduct of war. In thinking about the popularity of the major wars of the United States, I believe they historically rate as follows (No. 1 the most popular, No. 8 the least popular):

1. World War II
2. World War I
3. Spanish-American War
4. Korean War
5. American Revolution
6. War with Mexico
7. War of 1812
8. Vietnam War
9. [Civil War]

The Civil War has been bracketed and put last, because it obviously was a special case. Perhaps a historian ought to be cautious about commenting on current events, but one can guess that future historians will rate Vietnam as one of the most unpopular wars. What follows is a discussion of what were probably the two most unpopular foreign wars, save Vietnam. The War of 1812 and the War with Mexico present many interesting points for comparison and contrast.

First, a look at the War of 1812. Opposition could be found in all sections of the country to a degree, but in general New

England opposed the war, and the West, the South, and the Middle States supported it. Sentiment, however, divided along political as well as sectional lines. The Jeffersonian party everywhere supported the war; the Federalists, North, East, South, and West, opposed it.

Why was a war for maritime rights and for the ending of impressment supported by West and South, which had neither ships nor sailors, and opposed by New England, which had both? The answer is partly economic, partly ideological, and partly political. Actually the New England shipping interests did at one time object to the British Orders in Council and to impressment. But they objected to American embargoes even more. New England shippers were actually making money in the abnormal wartime carrying trade. The West and South were interested in normal trade—in the selling, not the carrying of goods.

Historians have shown that there was a depression in the West and South on the eve of war and that falling prices were blamed on England's maritime abuses. But Pennsylvania had no depression—and Pennsylvania staunchly supported the war. Why? Mainly, so far as we can tell, because she was Republican and supported the administration in general.

Many supported the war not only out of party loyalty but more importantly out of loyalty to the republican ideal. The United States was regarded by Americans—as well as by many republicans elsewhere—as the world's best hope, and the republican experiment was believed to be in peril.

But how extensive was the opposition to war? Henry Adams says the war was opposed by people who considered themselves, and were considered by their neighbors, as "not insubstantial"—in other words, people of property and place. Most vocal in opposing the war were Federalist politicians and the New England clergy. When war was declared Governor Caleb Strong of Massachusetts proclaimed a day of fasting and prayer. Not to be outdone by the Executive, the Legislature of Massachusetts issued an Address to the People advising them:

THE WAR OF 1812 AND THE MEXICAN WAR

Express your sentiments without fear, and let the sound of your disapprobation of this war be loud and deep. Let it be distinctly understood that in support of it your conformity to the requirements of law will be the result of principle and not of choice.

In New England, it must be remembered, there had always been a close connection between church and state. During the Revolutionary period the church had been disestablished in New York and all the states south. But the Congregational Church was still the established church in New England and was to remain so in Massachusetts until 1833. The New England clergy had always regarded the Jeffersonians with the deepest distrust. During the election of 1800 Jefferson had been denounced from the platform and pulpit. "Atheist, anti-Christ, Jacobin" were among the milder epithets bestowed. The Embargo, and finally the declaration of war against England, served only to confirm the Federalist preachers in their worst fears. Again and again the clergy charged that the Republicans had formed an alliance with that devil on earth, Napoleon.

The Reverend Timothy Dwight, who combined the role of clergyman with that of college president, summed up the case against the war for his students in the Yale College chapel:

1. A great part of our countrymen believe the war in which we are engaged to be unnecessary and unjust. . . .
2. We have begun this war, almost without preparation.
3. Our enemy is so situated as to be able seriously to disturb us. . . .
4. There is . . . reason to fear, that we may by this war be brought into alliance with France.

The Reverend J.S.J. Gardiner, rector of Trinity Church (Episcopal) Boston, told his congregation that: "It is a war unexampled in the history of the world; wantonly proclaimed on the most frivolous and groundless pretences."

When the prosecution of the war brought reverses and defeats, the clergy naturally saw the hand of God punishing a wicked people. God seems to have revealed himself to the New England clergy with a complete lack of ambiguity. The people of the South and West were, according to them, not merely unwise in their views; they were corrupt, tyrannical, and bloodthirsty. The Reverend Eliah Parish told the members of his congregation:

> You must in obstinate despair bow down your necks to the yoke, and with your African brethren drag the chains of Virginia despotism . . . those western states which have been violent for this abominable war of murder; those states which thirsted for blood. God has given them blood to drink. Their men have fallen. Their lamentations are deep and loud.

These quotations lend corroboration to the assertion that present-day opposition to war is not wholly new in the United States.

The next question is: how did this opposition affect the conduct of the war? Put very simply: the opposition deprived the country of victory and brought it to the brink of disaster. More particularly, the opposition had a direct affect on strategy, enlistments, and finance. To an audience composed of Army officers, it is hardly necessary to enlarge upon the importance of strategy, manpower, and money for the successful achievement of war aims.

The strategy of the War of 1812 has often been criticized—and criticized without reference to the cold realities of available support for and opposition to the war. The problem of taking Canada has often been compared to the hewing of a tree. The trunk of this tree was the main line of settlements extending along the St. Lawrence River from Québec to Montréal. The routes were the sea-lanes extending back to England. The branches were the settlements along the rivers and lakes in Upper Canada. To hew a tree one cuts it as close to the roots as possible; sever the trunk and all else falls. Important also was naval control of the Great Lakes, which formed the British line of communication westward.

THE WAR OF 1812 AND THE MEXICAN WAR

Now the fundamentals of this situation were well understood. No responsible leader, from the President on down, failed to grasp the importance of Québec and Montréal and naval control of the Lakes. Why then did the government attack the branches instead of the trunk? The answer lies in the opposition to the war. Those who opposed the war were right in saying that the country was unprepared. Though the War Hawks had talked of war for two years or more, they had done precious little to prepare the country to fight. The anti-military, anti-naval dogmas of the Republican party died hard. When the war was declared the authorized strength of the Army was 35,000, but only 11,700 men (including 5,000 recent recruits) were actually enrolled. These untrained troops were widely scattered throughout the country in various posts. The declaration of war came in June, 1812. If a campaign against Canada was to be undertaken before the winter set in, it had to be organized hastily. New England had the best militia in the country, but New England was also the place of the greatest opposition to the war. President James Madison favored striking against Montréal and "thus at one stroke [to] have secured the upper province, and cut off the sap that nourished Indian hostilities." The only way to take Montréal in 1812 was to call out the New England militia.

Under the Constitution the militia can be called into the service of the national government to execute the laws of the United States, to suppress insurrection, or to repel invasion. When asked to call out the militia of Massachusetts, Governor Strong, whose views have already been noted, told the President that he, as governor of the state, rather than the President, had the right to decide when the constitutional exigencies obtained. The requisition of the President was refused. The governor of Connecticut took a similar view and would furnish no troops. The constitutional question thus raised was not settled for the duration of the war. In fact, it was not settled until 1827, when the Supreme Court, in the case of *Martin* vs. *Mott*, decided that it belongs exclusively to the President to judge when the exigency arises to call for the militia.

Without the support of New England, a main thrust at the

trunk of the British-Canadian tree was not feasible. Hence something less than an ideal strategy had to be adopted. Not wishing to lose "the unanimity and ardor of Kentucky and Ohio," the President put into operation initiatives from Detroit, Niagara, and Sackett's Harbor. None of these was successful, and Brigadier General William Hull's attempt to invade Canada from Detroit ended in disaster. There is not space to follow the influence of opposition to the war on strategy in detail. Suffice it to say that the opposition and its geographical location prevented the government from carrying out the only strategy that could have brought the war to a successful and early conclusion.

Let us turn now to a related problem—that of manpower. As already indicated the best source of trained, or at least semi-trained, manpower was in the New England states. And lest a false impression be created, it should be pointed out that despite the opposition of some state governors, Federalist politicians, and the clergy, many patriotic men from New England enlisted in the Regular Army. In fact, some of the best regiments were recruited in New England. It was the Massachusetts 9th and 21st, the Vermont 11th, and the Connecticut 25th that did some of the best fighting at Chippewa, Lundy's Lane, and Fort Erie. In the number of recruits furnished the Regular Army, Massachusetts was second only to New York.

Even so the national government could not raise an army of a size sufficient to the needs. The main reliance was on the militia, but these were generally enlisted for active service of only six months. Enlistment in the Regular service was for five years, with pay at $5 per month, $16 bounty, and on discharge three months' pay and 160 acres of land. Volunteer organizations, officered under state laws, to serve for one year, could be accepted to the limit of 50,000 men. None of these methods or inducements sufficed to raise a viable army, and by the autumn of 1814 the manpower problem reached the crisis stage. On paper an army of 62,000 men was authorized, but actual strength in September, 1814 was only 38,000. In December, the land bounty was doubled to 320 acres; but even so, by February, 1815 the Army had dwindled to 33,000 men.

THE WAR OF 1812 AND THE MEXICAN WAR

In these desperate straits James Monroe, who in 1814-1815 was serving both as Secretary of State and Secretary of War, suggested plans for conscription. But Congress could not bring itself to adopt such an extreme measure. One may fault Congress for its timidity, but it must be recognized that the social and political conditions of the country imposed limitations. No one supported the war more ardently than Jefferson, but he wrote: "It is nonsense to talk of regulars. They are not to be had among a people so easy and happy at home as ours. We might as well rely on calling down an army of angels from heaven." In other words, it would have been difficult to raise an army for a popular war; for an unpopular war, the task was hopeless.

Now let us move to finance. Those who have any knowledge of the American Revolution will recall that one of the main difficulties was the inability to collect taxes. The Continental Congress had the power only to request money from the states. The Constitution adopted in 1789 provided the Congress with ample power to tax, but it is one thing to have a formal power and quite another to have the will and nerve to exercise that power.

In matters of finance, as in military matters, the Jeffersonians were hampered by doctrine. They had opposed the creation of Hamilton's Bank of the United States, and when its twenty-year charter came up for renewal in 1809, refused to recharter. The absence of a central banking authority and the chartering of unsound state institutions added to the financial difficulties.

It was New England that had the best sources of troops; likewise it was New England that had the best sources of money. The New England clergy denounced the war on moral grounds, but moral considerations did not prevent New Englanders from lining their pockets. New England shippers and bankers profited by legitimate trade and illegitimate trade, by operating in cooperation with British commerce and by preying upon that commerce, by smuggling, by carrying British goods, and by developing manufacturers to compete with British goods. In Massachusetts alone, specie holdings jumped from $1,709,000 in

June, 1811 to $7,326,000 in June, 1814. The New England Federalists hoped to bring about peace by withholding manpower and financial support. Obviously they were in a good position to do both. Harrison Gray Otis, a Boston aristocrat, was a director of several banks. His biographer, Samuel Eliot Morison, says: "Otis' correspondence gives indubitable proof that an excellent understanding existed between the financial powers of both cities [Boston and Philadelphia] to withhold subscriptions to government loans until peace was assured."

Alexander James Dallas became Secretary of the Treasury in October, 1814. He soon pointed out that not even the interest on the public debt had been paid punctually, and that a large amount of Treasury notes had been dishonored. "The hope of preventing further injury and reproach in transacting the business of the Treasury is too visionary to offend a moment's consolation," he said. In short, the government ended the financial crisis by ending the war.

One might well ask how it was that with so much opposition to the war, directly affecting strategy, manpower, and finance, the United States still managed to win. The answer is, it did not win. Militarily the War of 1812 was a draw. Neither side was able successfully to carry out a major campaign against the other. The Peace of Ghent, signed December 24, 1814, was a peace without victory. That the United States suffered no territorial losses was more the result of the international situation than of her own exertions.

Let us turn briefly now to the War with Mexico. As in the War of 1812, the opposition to the war was concentrated in New England. The War against Mexico, like the war against Vietnam, was denounced as an imperialist war to gain territory. It was portrayed as a conspiracy on the part of the slave states to gain more territory for slavery. Typical of this point of view was James Russell Lowell's *Bigelow Papers*. "Hosea Bigelow" is cast as an uneducated but crafty New Englander who speaks his mind in verse. The following stanza gives the flavor and direction of Hosea's opinions:

THE WAR OF 1812 AND THE MEXICAN WAR

> They may talk o' Freedom's airy
> Tell they're pupple in the face,
> It's a grand gret cemetary
> Fer the barthrights of our race;
> They jest want this Californy
> So's to lug new slave-states in
> To abuse ye, an' to scorn ye,
> An' to plunder ye like sin.

There is vast literature on the causes of the war: nevertheless, to assert as Lowell does, and as some modern historians do, that the war was a pro-slave plot is simplistic. At least two facts militate against this thesis. Some leading Southerners, including John C. Calhoun, opposed the war. Calhoun once said, "Mexico is forbidden fruit to eat of which is to die." His opposition was based on the ground that the territories that might be acquired, New Mexico and California, were not suitable for slavery. We should bear in mind also that the war was supported as much by the West as by the South. The Mexican provinces stood in the way of the westward-marching frontier. Manifest Destiny was a concept shared by many Southerners but not confined to them.

Just as in the case of the War of 1812, opposition was along party as well as sectional lines. The Democrats under James K. Polk had just fought and won an election by advocating a program of territorial expansion. A leading Whig editor and war opponent, Horace Greeley, wrote:

> People of the United States, your rulers are precipitating you into a fathomless abyss of crime and calumny. Why sleep you thoughtlessly on its verge as though this were not your business, or murder could be hid from the sight of God by a few flimsy flags called banners. Awake and arrest the works of butchery 'ere it shall be too late to preserve your soul from the guilt of wholesale slaughter.

How did the opposition affect the prosecution of the war? In this instance, dissent and criticism actually had very little effect on military operations. The results of the war produced profound political and moral divisions, but opposition during the war did not greatly affect the conduct of the war, begging the question, why?

The manpower and financial situations should be examined as in the case of the War of 1812. In the War of 1812 the fact that the opposition was concentrated in New England prevented the government from successfully invading Canada along the Lake Champlain route in 1812. In the War with Mexico the situation was reversed. Those who supported the war and wanted to do the fighting were relatively close to the scene of operations; and they could be, and were, called out. During the Mexican War troops were raised through volunteering. President Polk sensibly asked those states that had voted for him to supply volunteers. There was no great difficulty in raising troops.

So far as finance is concerned the country generally was much richer than in 1812, and the wealth was more evenly spread through the various sections of the country. The international situation, which those who study war ought never to ignore, was more favorable. The potato famine in Europe created a tremendous demand for American agricultural products. The prosperity of the country grew as the war progressed.

President Polk was also far superior to President Madison as a Commander in Chief. Even among the war's supporters there was much dissension about how to fight it, in Polk's Cabinet itself as well as elsewhere in the Federal government. But Polk was a man of great determination, and he had very clear ideas about what he wanted to do. Very likely one could read everything in the library and not receive much enlightenment on this point. No one who has written on Polk has understood two basic factors: the military situation under which he operated, and the concept of limited war.

The military system at the time of President Polk was a mixed one: partly political and partly professional, but the political was the dominant ingredient. All the high-ranking officers—one thinks particularly of Winfield Scott and Zachary

THE WAR OF 1812 AND THE MEXICAN WAR

Taylor—were political generals. All regarded their military careers as a means of rapidly rising in the politico-military sphere. For examples they had Washington, Jackson, and William Henry Harrison. Polk meant to be the politician in chief as well as the Commander in Chief. He simply *had* to play off one political general against another. Any other *modus operandi* would have placed him at the mercy of his subordinates.

It is unlikely that President Polk had read Clausewitz, or indeed that he ever even heard of the great German writer. But in any case Polk planned and carried out the strategy of the Mexican War to conform closely to Clausewitz's admonitions on the subject of limited war. Clausewitz, and some of the later theorists, said that a limited war should have concrete, realizable, negotiable aims; it should be limited geographically and in manpower and resources; civilian life should proceed pretty much as usual; and finally, one should seize the objectives one wants and try to persuade the enemy it is not worth the effort to recover them.

Polk's conduct of the war cannot be followed in detail here, but we can note briefly some of the highlights. The objectives of the war were to fix the boundary of Texas at the Rio Grande and to annex New Mexico and California to the United States. In order to achieve the first aim, General Taylor was stationed along the border, and when war broke out several battles were fought which put the United States in military possession of the disputed territory north of the Rio Grande. Colonel Stephen Watts Kearny marched southwest from Fort Leavenworth and took Santa Fe, New Mexico. His small American force was then divided three ways: a few troops were left in Santa Fe to garrison that post; another small detachment was sent into the northern provinces of Mexico; and with the remainder of his forces Kearny marched to California. There, with the aid of native Americans and the United States Navy, California was secured. By March, 1847, the United States had military possession of Mexico's northern provinces.

Having secured what was wanted, the United States hoped

to convince Mexico it would not be worth her while to retake the lost provinces. Mexico could not be convinced. So what was the last move? The last move consisted of an amphibious landing at Veracruz to establish a base, a march inland, and the taking of Mexico City. All this was done with civilian life in the United States going on pretty much as usual. Once Mexico City was taken there arose a hue and cry to annex the whole of Mexico. Manifest Destiny threatened for a while to sweep the country. But Polk stuck to his original limited goals.

In contrast to the War of 1812, Polk was served by able generals who commanded troops of generally high morale, though they were still poorly disciplined. It is hard to argue with success. Opposition there was, but it did not have a chance to interfere seriously in a short and successful war. Had the war dragged out, and if television cameras had been on the scene to record some of the excesses, it might have developed into another Vietnam. Again, to say that the opposition had little effect on the conduct of the war is a very different thing from saying that the conduct and results of the war itself did not have important political consequences.

THE WAR OF 1812 AND THE MEXICAN WAR

Bibliography

In my book *The War of 1812* (Chicago: University of Chicago Press, 1965), I have an analytical discussion of sources under "Suggested Readings." Since the appearance of my book, Reginald Horsman has published *The War of 1812* (New York: Knopf, 1969), which contains a "Bibliographical Note" especially useful for British sources currently available.

In preparing this lecture I have relied particularly on the following: Henry Adams, *History of the United States in the Administration of Jefferson and Madison, 1801–1817* 9 vols., (New York: Charles Scribner's Sons, 1889-1891); Charles L. Dufour, *The Mexican War: A Compact History, 1846–1848* (New York: Hawthorn, 1968); Samuel Eliot Morison, Frederick Merk, and Frank Freidel, *Dissent in Three American Wars* (Cambridge: Harvard University Press, 1963); Samuel Eliot Morison, *Life and Letters of Harrison Gray Otis* 2 vols., Boston: Little, Brown, 1913); Otis Singletary, *The Mexican War* (Chicago: University of Chicago Press, 1960); Justin H. Smith, *The War with Mexico* 2 vols., (New York: Macmillan, 1919); Emory Upton, *The Military Policy of the United States from 1775* (Washington: Government Printing Office, 1904); T. Harry Williams, *Americans at War: The Development of the American Military System* (Baton Rouge: Louisiana State University Press, 1956).

COMMENTS

All wars tend to be unpopular wars; but it is also true that unconventional wars tend to be especially distasteful to the people and governments of the developed powers. The armed forces of those powers are designed primarily to combat similar armed forces in conventional war, according to accepted rules governing warfare that have developed among the Western powers over the centuries. When an opponent is not similarly equipped and trained, even when his equipment and training are on the face of it patently inferior, the armed forces of the great powers may find their own structures awkwardly irrelevant to the problems at hand; against ill-equipped adversaries who resort to unconventional, especially guerrilla, tactics and ignore the accepted rules of war, the very wealth of matériel and experience of a great-power army may get in the way of effective prosecution of the war at hand. So it has always been when great armies fight "small wars."

There are also more profound reasons why unconventional wars tend to be especially distasteful to the developed powers. When does a great power find itself obliged to fight an unconventional war? It is not merely borrowing Communist terminology, but stating something close to the essence of the answer, to say that such a situation develops when a great power casts itself in the role of an imperialist nation resisting a movement of national liberation. And that role is exceedingly unpleasant for a great power, because the history of the past several centuries suggests that if it is a genuine nationalist demand for self-determination that is being resisted, then permanent resistance is virtually impossible, no matter how great the material advantage of the forces opposing the nationalism. In the long run, genuine nationalist movements almost always break-free. Such seems to be the lesson of all efforts to subdue movements for national independence, from the American Revolution through all the wars of national liberation and unification of nineteenth-

COMMENTS

century Europe—the struggles of Greeks, Serbs, Italians, and so on—through the recent futile struggle of France to retain Algeria—to say nothing of the struggles of both France and the United States in Indochina. (To be sure, if genuine nationalism practically always succeeds in the long run, then the Southern Confederacy of 1861–1865 may have to be written off as something less than a real expression of nationalism; Confederate nationalism was always so ambivalent, the continued affection for the old Union within the South always so persistent, that such a verdict may well be correct.)

In the Boer War, the British Empire achieved one of the rare successes even for the short run of an imperialist power over a nationalist movement. This British success, albeit it proved something less than permanent, might seem to make the Boer War a candidate for special study among the great powers, in how to resist wars of nationalist self-determination waged by unconventional means. But the Boer War has received far less study than might be expected, even in Britain, let alone in the other Western powers where its details are virtually unknown. That this war should be so widely ignored in itself implies much about the distastefulness of a great power's waging an unconventional, counterinsurgency war. Of course, the British do not like to be reminded of the series of defeats with which the Boer War began for them even when the Boers were fighting relatively conventional campaigns—too conventional, probably, for their own good. But it is even more painful to recall the expedients to which the British Army resorted when it was desperate for success against an unconventional campaign of insurgency.

Thomas E. Griess in 1971 summed up what most readers will find an unfamiliar story; as a member of the history faculty of the United States Military Academy, he appropriately earned his Ph.D. in history at Duke University with a study of the Academy's first great teacher of the art of war, Dennis Hart Mahan.

A Case Study in Counterinsurgency: Kitchener and the Boers*

THOMAS E. GRIESS

Background

BEGINNING in October, 1899, and lasting for almost three years, the Boer War degenerated in its latter half into a struggle between guerrilla and conventional forces. By September, 1900, British troops under Lord Roberts had dispersed the Boer levies and occupied the chief towns in the Orange Free State and the Transvaal. With the seemingly complete collapse of Boer resistance, it appeared that the power of the two Boer republics had been irrevocably broken. President Paul Kruger of the Transvaal had fled to ultimate sanctuary in Holland in mid-September, and, although it was known that a few thousand apparently disorganized burghers had ridden off into the rugged hills of the northeastern Transvaal, the war appeared to be at an end. If there were reports of guerrilla activity in the Orange Free State, by late November Roberts could discount them and think it time to go home to a hero's welcome, leaving the relatively simple policing tasks to his chief of staff, Lord Kitchener. The British leaders who held this view had misread the signs.

The Boer republics were not in shambles, although their military forces were scattered, their capitals occupied, their railroads lost. Civil power resided in the persons of Presidents

---

*This essay appeared as Chapter III in LTC Veloy J. Varner, ed., *Development of Revolutionary Warfare: French Revolution to World War II* (West Point: United States Military Academy, 1970).

Marthinus Steyn of the Free State and Schalk Burgher of the Transvaal, with whom acknowledged military leaders maintained contact. A concerted plan for continued resistance to British arms had been formulated by these leaders. This resistance, however, would exploit guerrilla tactics. The Boers had decided to abandon the trappings of conventional warfare and to utilize "every blade of grass, every brown ironstone boulder or shelf of purple shale, every contour of the illimitable veldt, and every corner of every kopje. . . . What Roberts had won was a shadow. He left Kitchener to grapple with reality."[1]

The methods Kitchener employed to combat the new Boer tactics are illuminating to the student of counterinsurgency. Evolving slowly as the British gained experience during the year and a half remaining in the war, these methods gradually but inexorably reduced the Boers to capitulation. That they were militarily effective, considered as a whole, one can hardly question, for the Boers finally sought and accepted peace on British terms. Why the methods were effective and how they might have been improved is of far more importance. The British counterinsurgency effort is significant, for although history cannot foretell the future, it can provide a base of experience and a means of interpreting contemporary events. Common causes, problems, and solutions may exist. Providing one consciously avoids the pitfalls of anachronism and realistically weighs dissimilarities, such common factors may be fruitful subjects for study. A critical examination of Kitchener's methods can provide a framework for such a study. Before considering the explosive situation confronting Kitchener upon his assumption of command in November, 1900, however, a brief recounting of events leading up to that point is desirable.

Causes of the War

As with most major trials by arms, the causes of the Boer War have been the subject of controversy and debate. The

British government objected to the discrimination against British subjects in Boer constitutions that strictly forbade the non-Boer franchise. Furthermore, no person could hold municipal office in the Boer republics unless he was a member of the Dutch Reformed Church. British businessmen were also penalized by monopolistic regulations favoring the Boers. Finally, the Crown was disturbed about the open Boer preparations for war.[2]

The Boers, on the other hand, resented British interference in their internal affairs, rejected the concept of British suzerainty, and recalled that they had defeated the British regular forces in the First Boer War (1881). They were determined to remain independent. It was this goal, steadfastly adhered to by a handful of influential leaders, particularly in the Orange Free State, that sustained the Boers throughout the war. In January, 1900, Kruger telegraphed Steyn that "this war can only be ended in one of two manners: either by our practical extinction or by our getting what we want. With us, the only question is one of freedom or of death."[3] Fourteen months later, when some Transvaal leaders appeared to waver and Kitchener seemed amenable to peace, the fiery general in the Free State, Christiaan De Wet, brusquely refused to consider a settlement since nothing short of independence was acceptable to him. And on the eve of the peace, De Wet went amongst his subordinates urging them to vote for peace at the forthcoming Boer council only if independence was guaranteed.[4] Nor would Koos De la Rey, the unquestioned guerrilla leader in the western Transvaal, compromise Boer independence until the day before the council voted for peace in May, 1902.[5]

This tenacity of purpose was buttressed by another factor that helped sustain the guerrilla effort. The Boers distrusted, even hated, the British government, which they considered to have driven them unjustly into war. Their hatred found its maker in Lord Milner, the High Commissioner at Capetown, whom the Boers despised. His appointment and establishment of residence at Pretoria as Governor of the two new colonies, after Roberts' campaign and British annexation, merely heightened the provocation. The Boers had not forgotten his intransigent and haughty attitude at the Bloemfontein Conference in 1899, and they correctly suspected that his goal was nothing less than complete British domination and colonization of the Boer states. They feared the worst at the hands of Milner, who from the beginning of the war had been a staunch advocate of no terms except unconditional surrender.[6] Consequently, they continued to fight under ever-worsening conditions, in the forlorn hope that the British people would tire of the struggle, that the home government would topple and be succeeded by a Liberal ministry more friendly to them, or that foreign intervention—at least the pressure of foreign opinion—would become an effective third force.

The Conventional Phase of the War

Before they formally decided to wage guerilla warfare in September, 1900, the Boer leaders elected to fight a conventional war with the British. In the short-lived first phase of the struggle, when the Boers outnumbered their enemy two to one, they laid siege to Ladysmith, Kimberley, and Mafeking instead of fanning out through Natal and the Cape Colony in maximum strength to enlist support from the Cape Dutch and possibly seize control of areas vital to the coming British efforts at reinforcement. A few young Boers, such as Louis Botha and Jan Smuts, saw the opportunity, but they were too lacking in influence to mold policy.[7] Their alternative strategy, however, remained a shining hope for these same leaders for the rest of the war. On several occasions, Boer columns penetrated into Cape Colony, primarily in attempts to spark mass uprisings. The Cape became a fixation as the possibility of foreign intervention dwindled; in desperation the Boers sought support there, months after the one chance for success in this respect had gone glimmering. The second phase of the war began with Roberts' arrival in January, 1900, and ended in November, with the republics supposedly ground into submission.

KITCHENER AND THE BOERS
The Boer Insurgency

During the second phase of the war the situation developed in a manner favorable to guerrilla tactics. The Boers were united in detesting the British, thus providing a base of support for the guerrillas. Some of the more fainthearted burghers and townspeople had surrendered, but most of the former group had merely returned to their farms to await calls to action. And the really hard core of patriots, with its handful of dedicated and skillful leaders, had melted into the hills whence it would emerge to galvanize the remainder into action.[8] The theater of operations was vast and, provided he was mobile, it afforded an ideal arena for the irregular. Almost destitute of trees except along the river banks, it was generally flat. There were, however, occasional ranges of hills and isolated *kopjes*, and in the east and southeast there rose imposing mountains. Drinking water was scarce.[9] Guerrilla tactics came naturally to the Boer, raised as he had been in a relative wilderness, taught to rely upon his skill with a rifle, trained in the use of ambush, and appreciative of the value of surprise. The British troop units, on the other hand, were dependent upon fixed bases and miles of vital railroad serving them; both presented ideal targets to the ubiquitous Boer. Finally, the Transvaal government had managed to spirit a supply of gold from Pretoria before Roberts seized the capital; this sum provided initial financial backing for a guerrilla war. These conditions favorable to irregular warfare, however, cannot alone explain the Boer successes; they also had to have a form of organization as well as a tactical system.

Resentful of what they viewed as British infringement of their sovereignty, the Boer republics had begun procuring arms in large quantities by 1894. When war came the Transvaal probably had between 60,000 and 80,000 modern rifles, at least half of them German Mausers which used smokeless powder cartridges employed in clips of five rounds each. The Free State armed less vigorously, it being agreed that Kruger's republic would arm for both. There were also on hand at least one million rifle cartridges

333

and about a hundred modern artillery pieces of varying calibers manufactured by Krupp and Creusot.[10] By the time the Boers began active guerrilla operations, however, much of this material had been lost or deliberately destroyed. They retained only a few artillery pieces, and before many months elapsed they abandoned most of these in the interest of mobility. Nor did the supply of rifles last, even though excess stockage was secreted upon movement into the hills in September, 1900. Too much credit for Boer success, accordingly, should not be given the superior Mauser rifle, since in the final year of the war the Boers frequently utilized weapons lost by or captured from British troops.[11]

The Boer Military System

The districts into which the Boer republics were divided served as the building blocks for the military organization. Each district, dependent upon population, furnished a commando of from 300 to 3,000 men. The burghers in each commando then elected a commandant to lead them in battle. In supreme command was a commandant general who was appointed by the state; Louis Botha and Christiaan De Wet occupied these positions for the Transvaal and Free State, respectively, during the guerrilla phase of the war. Within the commando, the commandant was assisted by field cornets and corporals who were also elected by the burghers. These field cornets, coming from the wards into which districts were divided, had civil duties in peacetime (e.g., to inspect natives, to serve as justices of the peace) and were key individuals during mobilization for war when all males between the ages of sixteen and twenty were marshalled for service.[12]

A distinctive feature of the Boer military system was the great degree of independence the individual exercised. The officer could not order men to take part in proposed operations—he asked for volunteers. Strategy was decided at councils of war (*Krijgsraad*), ostensibly restricted to the leaders but always attended by many of the burghers in camp, who aired their views.

Furthermore, the Boer felt no compunction about leaving the *laager* to return home to be with his family or even to retreat in battle without orders if it appeared that he was in danger of being surrounded. It is generally true, however, that as the guerrilla war intensified, aggressive leaders exercised increasing control over operations and imposed a greater degree of discipline upon their burghers. Commandant Ben Viljoen, one of the key leaders in the Transvaal, constantly had problems of insubordination and deplored "the Boer's exaggerated notions of freedom of action and speech." De Wet, too, found this tendency discouraging and remarked to General P.J. Joubert after the fall of Bloemfontein that "whatever I had said or done the burghers would have gone home."[13]

This spirit of independence came naturally to the Boer, raised as he had been in a vast wilderness where at an early age he had developed self-reliance and practical skills as an individual fighter. Moreover, it contributed in no small measure to his superb qualities as a guerrilla fighter; for in actions involving small groups of irregulars, individual initiative, skill, and intelligence played important roles—every man his own general, so to speak. It was less of an asset when the Boers attempted to fight stand-up battles in the tradition of formal warfare. The decision to wage guerrilla warfare recognized and exploited this national attribute; Botha, in particular, was careful to disperse his commandos to their home districts where the burghers could easily satisfy their desires for visits home and where they would have more incentive for fighting. General Redvers Buller, although during the war he had not shown any great understanding of Boer tactics, seems to have appreciated this point; testifying in 1903 before a Royal Commission investigating the war, he observed that "to every man his own home is the capital."[14]

Boer Tactics and Military Assets

The Boer system of tactics was also peculiarly well adapted to guerrilla operations. One historian has claimed that the Boers were not a people of civilians at all, " . . . but a fighting race

UNPOPULAR AND UNCONVENTIONAL WARS

with a fighting history."[15] Early in his history the Boer had learned that his survival depended upon personal skill and a trained horse. He fought the British as he had learned to fight natives and beasts, relying upon skillful use of the rifle and avoiding hand-to-hand combat. His objective was to get no closer to the enemy than necessary for accurate rifle fire, and to use the ground carefully to permit getting within range but behind cover and unobserved. He utilized an extended formation that emphasized flanking positions and depended upon firepower to cover gaps created by the extension. His horse remained in a protected position behind him, trained to stay there without being guarded; it was readily available for quick lateral movement to reinforce sections in trouble or for quick withdrawal when the action threatened the Boer's survival. He preferred defensive positions. The question was "not how to attack [the enemy], but how to find a position suitable for repulsing his attack,"[16] as well as one from which he could deliver heavy, accurate fire, withheld until the last possible minute. Only rarely during the guerrilla phase of the war did he make mass attacks or cavalry charges—and when he did he usually suffered defeat.[17]

As a marksman the Boer was highly competent, although he has probably been overrated. Without question, however, the father of the Boer soldier of 1900, practiced in hunting game and often fighting natives, had a skill with the rifle that was phenomenal. In an engagement against the British in 1881 the Boers inflicted 66 percent casualties in ten minutes, the wounded averaging five wounds per man.[18] Concerning the marksmanship of the Boer in 1900, the opinion of veteran leaders of the British Army was mixed. General J.C. Ardaugh testified that by 1899 game was no longer plentiful and that the Boer, though a better shot than the Briton, was not as good as the Australian. General J.P. Brabazon observed that the Boer did his damage at "close" ranges—less than 300 yards—but was not very effective beyond that distance. General Charles Knox, however, asserted that the Boers hardly ever shot without "hitting something or the other.'"[19] We may conclude, therefore, that the Boer could generally outshoot his British counterpart, who had a healthy respect for his marksmanship.

The mobility of the Boer insurgents was probably the key factor in their success as guerrillas. Their horses were hardier than those of the British and could absorb an astounding amount of punishment. These amazing ponies were able to travel sixty miles a day for several days in succession and were capable of speeds beyond that of the animals used by the British. Furthermore, they could operate as much as three days without food.[20] Since the average guerrilla had more than one of these horses, he could usually rely upon a relatively fresh mount. Nor did he excessively load the mount, seldom carrying more than his rifle, ammunition, and a bag of dried beef. He could travel lightly because provisions could be obtained from the local farms and because he meticulously conserved his ammunition. When several commandos assembled for a lengthy operation, a wagon train containing supplies and ammunition was usually made up for the use of the group; some leaders, however, even refused to take these convoys with them. Such trains were frequently captured by pursuing British columns because the oxen were too slow. When this happened the Boers merely assembled another train from wagons and supplies in the friendly countryside during a period of recuperation.[21] Finally, the Boer was thoroughly familiar with the country, and even when he was closely pursued he was able to employ feints and guile to throw his opponent off the track. This, then, was the formidable irregular, operating as a member of an organized guerrilla band, who faced Kitchener upon his assumption of command in November, 1900.

British Counterinsurgency

The new British commander could hardly fail to realize that the war was far from finished and that to cope with the Boers he required a different type of army and new methods. As early as August, De Wet had returned to the Free State and had begun whipping up the fires of resistance among the burghers. With most of the British Army far away in the Transvaal, he intensified the raids on railroads and isolated garrisons. By the first of the year, the Transvaal Boers also had made the transition from formal to guerrilla warfare and were actively engaged in raiding. By June, 1901, the Boers had cut the 1,300 miles of railroad more than 250 times in the past year, and Royal Engineers had rebuilt some 225 bridges and culverts that raiders had destroyed.[22] To bring De Wet to bay, Roberts had sent troops back to the Free State and organized columns to comb the countryside. These so-called mobile columns, however, were composed primarily of infantry, heavily burdened with immense ox-drawn supply trains. The Boers politely got out of their way and then reoccupied localities after they had departed. A captain in one of the British mounted units commented that these columns had "as much chance of catching Boers on the veldt as a Lord Mayor's procession would have of catching a highwayman on Hounslow Heath. . . ."[23] Of this basic problem Kitchener was well aware. He also appreciated that it would take time to effect a remedy. An urgent request to London asked for more mounted troops and replacement horses. At the same time, Kitchener energetically set about revitalizing troop units, shifted them about, and, for a lack of better means, continued the slow-moving countryside drives, for the Boers were dangerously close to seizing the initiative. But he also tried another approach.

Negotiation

Kitchener enlisted the aid of surrendered burghers in trying to convince the Boers of the hopelessness of continuing the war. The commanding general was a practical man, and the reader might logically conclude that, knowing the success the commandos were having, he could not honestly expect them to sue for peace—that he was merely trying to buy time until his army was reorganized to deal more effectively with the guerrillas. Perhaps this was a part of his objective. At the same time, however, he also favored negotiating with the outlawed governments and objected to Milner's stand on unconditional surrender. If the Burgher Peace Committee could at least have brought about a conference with Botha, Kitchener would have been delighted. Furthermore, he was receiving complaints from surrendered burghers—"hand-uppers" the guerrillas disdainfully called them —that the commando forces were exerting pressure upon them to serve with the guerrillas. There is no question that this was being done, and with considerable success. Botha had advised one of his subordinates in October to promulgate the word that if the burghers did not stop laying down their arms he would "be compelled, if they do not listen . . . to confiscate everything movable and also to burn their farms."[24] Thus, Kitchener encouraged the Burgher Peace Committee to send emissaries into the commando camps. The Boers were genuinely worried about the effect these efforts would have on the commandos and treated the emissaries harshly, going to the extreme of trying and executing as a traitor the leader of the committee, Meyer De Cock.[25] And so the organized efforts of the committee came to nought, though one recommendation it made to Kitchener had far-reaching consequences.

Attempt to Isolate Guerrillas from Population

Taking the advice of the Peace Committee, the British commander decided in December, 1900 to exert pressure on the Boers by evacuating into camps under British control all inhabitants in areas where the Boers were active.[26] A few camps already existed at this time for refugees who had sought protection from the guerrillas. Kitchener's object, of course, was to intimidate the Boers and strike at their bases of supply. In fairness to him it must be noted that he was also moved by humane considerations and so stated in his order, for some of the people affected were almost destitute in consequence of Roberts' earlier order that all farms belong to Boer supporters would be burned and the livestock destroyed. Kitchener later justified his evacuation proclamation on the ground that Botha frankly told him at their Middelburg conference in March, 1901 that he would not cease forcing men to join his forces and, if they refused, destroying their property, thus leaving destitute families on the *veldt*.[27]

The consequences of the evacuation policy were far-reaching. Kitchener had not foreseen the magnitude of the housing, medical, sanitary, and feeding problems. The Army, more concerned with military missions and short of manpower, provided inadequate staffs for the camps and logically assigned the less efficient personnel to run them. Conditions deteriorated, many of the internees died, a cry of indignation came from pro-Boer elements in England, and the Boers gained sympathy throughout the world. The government took control of the camps out of Kitchener's hands and turned it over to Milner, who gradually corrected the evils. By the time the war ended, the Boer leaders were acknowledging with thanks that their dependents were under British care. Those dependents in Boer *laagers*—Kitchener ceased evacuating them in the final months of the war—were in a pitiful state; when asked what to do with them when he wanted to continue the war, De Wet advised sending them to the British in the company of a token group of guerrillas who would surrender.[28] The pertinent question, as far as this study is concerned, however, is how effective the evacuation policy was as a counterinsurgency measure.

The resolution of the evacuation question is closely associated with the policy of burning farms, since removal of families to a place of safety inevitably followed the policy of devastation. As this devastation became widespread and was directed by an increasing number of Kitchener's subordinates, there came to be little discrimination between property belonging to "hands-uppers" and to commando sympathizers. Hence, burghers who might initially have favored peace could see their avowed protectors destroying their property, and consequently they might well swing over to the side of the guerrillas. There is also the argument that the destruction of property worked to the disadvantage of the British in that it strengthened the resolve of the Boer guerrillas to continue resistance.[29] In view of their adamantine attitude toward independence, however, it is unlikely that the Boers would have been inclined to surrender earlier had Kitchener left the farms undisturbed as Botha wanted him to do. These were an important source of supplies for the guerrillas, and Kitchener could hardly ignore them.

There are additional arguments against the British policy of destruction of property. John F. C. Fuller maintained that leaving the farms undisturbed and the women thereupon would have created traps which British troops could spring.[30] There are two fallacies in this argument: first, the Boers were too skillful and cunning in contrast to the *average* British unit, and, second, the British had too few troops to set such traps. General Charles Knox, an expert in trying to cope with De Wet, testified that many Free State burghers would have been content to stay on their farms if assured of protection from De Wet and his militants, but that the British never had enough mounted troops to deal with the scattered bands of Boers.[31] A further consideration is that the policies of devastation and evacuation distracted the troops from the real mission of defeating the guerrillas.[32] This is a just charge, but it is worth noting that when the devastation policy was being more vigorously implemented, the British forces were not yet so organized as to be capable of catching the Boers. At the same time, the destruction of supplies was contributing, however slowly, toward sapping the ability of the Boers to fight. Based upon the testimony of their leaders when final peace negotiations took place, there can be little doubt that their inability to obtain supplies had finally become an instrumental factor in bringing about their capitulation.[33]

The major criticism of the policy of evacuation of Boer dependents to camps is that this freed the guerrillas from responsibility for their families and allowed them to devote their full attention to the war. Theoretically, this stands as a logical conclusion, and it would appear to be sustained by De Wet's solution, noted earlier, for continuing the war. But it may overlook the humane factor, the very same factor that, applied to the camps, brought so much criticism of the British. If it is accepted that the bases of supply had to be destroyed, could Kitchener have left women and children on the *veldt* to starve? Admittedly, he did this late in the war, but by then the numbers were small. Such a policy would have thrown the burden of care upon the Boers, who would likely have been driven to capitulation earlier than they were. Their indomitable spirit, however, would have universally blamed the British. Another factor, considering the closeness of family ties in the Boer civilization, is that it is unlikely that the guerrillas were completely devoid of concern for those of their dependents who were interned by the British.[34] This concern would sap the will of some guerrillas to resist, although inadequate evidence exists to sustain this claim for the guerrillas as a whole. In conclusion, the internment of Boer dependents was detrimental to the overall British counterinsurgency efforts, but it exerted an influence toward peace upon some individual burghers.

In a vein similar to his internment edict, Kitchener attempted two other policies of intimidation that were relatively ineffective. Plagued by Boer incursions into the Cape Colony and Natal[35] and sporadic uprisings by Boer sympathizers therein, Kitchener proclaimed martial law and took stern measures against captured rebels. This enabled him to employ military measures in what normally would have been rear areas of operation. His measures offended the two colonial governments, and the execution of a few rebels, sometimes done too flamboyantly, did not cow the burghers friendly to the Boers. When the war ended, operations in the Cape were at a stalemate; there were 3,000 Boers and rebels therein, but a general uprising was not in prospect. To bring about this balance, Kitchener had been forced to divert troop units to the area.[36] As a counterinsurgency measure, one cannot quarrel with Kitchener's decision on martial law, though there is a lesson here in civil-military relations. Both Roberts and Kitchener, however, might have attempted to organize and use more effectively loyal troops from the two colonies to cope with Boer raiders, thereby conserving their front-line units for employment in the Free State and the Transvaal.

The second measure of intimidation was Kitchener's proclamation in August, 1901, threatening banishment for life from South Africa for Boer leaders unless they surrendered. De Wet's reaction that "Nobody dies of fright" was typical of the disdain with which the Boer higher echelon greeted the announcement.[37] This attempt at intimidation, Kitchener's last effort to bluff the Boers into surrender, came at a time when he was experimenting with new military techniques for use against the guerrillas.

The Military Solution

Kitchener's military solution for coping with the ubiquitous Boers evolved over a period of several months and consisted of a combination of three features: blockhouses, drives by mobile columns, and surprise night raids. A fourth feature, centralization of control, was concomitant to the first two. A stickler for detailed planning and precise execution, the British commander concluded that the best chance of defeating the Boers lay in his personally controlling by telegram a number of mobile columns which he could manipulate in response to intelligence assembled at his headquarters. Ultimately this system achieved results, but it may not have been the best system. Kitchener's approach assumed that the British troops and leaders were not the equal of the Boers. In general this was a correct assumption, but the system discouraged his subordinate leaders from exercising initiative and developing unorthodox counterstrategies; it made them slaves to a scheme. Furthermore, Kitchener's sifted intelligence tended to be more stale than that which his few sector commanders could have collected with organizations oriented on their respective sectors. There was a lower limit to which responsibility for operations could be delegated; but in dealing with the highly decentralized and wily Boers, more decentralization of British command and control was desirable.

Drives by Mobile Columns

Kitchener's first great drives were initiated in January, 1901, before the additional mounted troops he had requested arrived.[38] By this time the British organization had clearly substituted the column for the division as the common military unit. These columns were designed to be as mobile as possible, but they still included sizable infantry groups and supply trains. Not until May, when Kitchener had finally assembled about 80,000 mounted troops,[39] did they begin to attain the mobility necessary to compete with the Boers. Numbers alone, however, were not enough; training and experience had to follow. For the time being Kitchener had to employ the troops he had immediately available. The drive in the Orange Free State against DeWet, made by approximately 15,000 men in fifteen columns, employed most of the experienced mounted units. In this operation the British hounded DeWet with one column in direct pursuit and, using the railroads for rapid realignment, placed others in blocking positions. But the British columns were neither mobile nor experienced enough, intelligence was tardy in arriving, and DeWet escaped after exhausting his pursuers. The other drive, in the eastern Transvaal, was designed as much to exploit the new policy of devastation of the countryside as to pursue Botha. Twenty-one thousand troops in seven columns were employed; mounted units were few, and movement was methodical and slow. Botha had no trouble eluding the columns, but the countryside suffered its first major systematic rape.

Though for the rest of the year Kitchener continued to mount drives of varying size and in the Free State used increasingly mobile columns, results were frustrating. Farms and cattle were destroyed, wagons captured, families swept into the camps, but few prisoners were taken from the hard core of the guerrillas. Nevertheless, the troops were gaining experience as the tactics of the drive were being refined. In 1903, General M.F. Rimington, one of the more imaginative column commanders, explained how he had tried to improve the efficiency of his force: lighten the load carried by the horse, make soldiers dismount when halted to rest horses, allow horses to graze whenever possible, scout at night to locate Boer *laagers*.[40] (During the last year of the war his column of 2,500 mounted troops covered about 3,500 miles, suffered 55 casualties, lost 1,617 horses, killed 45 Boers, and captured 396 Boers, 1,800 horses, 391 wagons, and 30,000 head of cattle.)[41] The two major reasons the drive was not capturing more Boers were: first, the columns could not cover the ground at night adequately enough to prevent the quarry from slipping between them, and, second, the centralization of control under Kitchener prevented column commanders from acting instantly in response to changed situations. Thus a drastic change was bound to come.[42] It was Rimington who suggested to Kitchener the final refinement in the drive technique: the use of a continuous line of horsemen, its flanks resting on blockhouse lines strengthened by infantry, riding straight forward by day and halting at night to form a solid entrenched line. Kitchener aimed to form a net and flush the Boer into it.[43]

The British used the new technique in early 1902 in two drives against De Wet. Kitchener was not able to catch the wily Boer, who broke through the line at night, but he captured 1,100 of his men, several wagons, and a few dependents. The thin line in the new-model drive could be broken by a determined enemy willing to take moderate casualties; and it depended too optimistically upon alertness at night along a very long perimeter— 9,000 men were in a cordon fifty-four miles long in the first drive.[44] But it was the most effective measure yet tried; its grinding methodicalness struck at Boer morale and harried them into desperation dashes to and fro which wore out horses; and, of paramount importance, it demonstrated the fantastic extreme to which Kitchener would go and how implacably determined he was. As events developed it became unnecessary to continue the drives. Under the cumulative weight of British pressure, the Boers negotiated for peace in May, Kitchener continuing operations to the very last moment in the western Transvaal as an ominous warning. This narrative, however, has proceeded beyond the development of the blockhouse, the most significant of any of the British innovations.

Blockhouses and Barriers

Starting in January, 1901, Kitchener constructed long chains of blockhouses linked by barbed-wire entanglements that prevented the Boers from moving freely about the country.[45] Raiders could not cross a blockhouse line without being seen and without coming under fire from at least one blockhouse. As first applied, the system was a purely defensive measure, almost one of desperation—to protect the railroads and to render a few key towns secure. Located near large railroad bridges, the initial blockhouses were two-story masonry buildings, expensive to build in terms of manpower, time, and materials.

Seeking a more economical substitute, Major William Rice, Royal Engineers, developed an inexpensive, portable blockhouse in February. Since the Boers no longer had artillery pieces, he designed an octagonal shelter with inner and outer walls of corrugated iron four and one-half inches apart, the intervening space being filled with hard shingle and thereby capable of withstanding rifle fire. The sheeting was nailed to a simple wooden frame that rested on a level earth foundation: a simple overhanging roof was added. Further experimentation produced a circular shelter, consisting basically of two iron cylinders of different diameters forming a protected space thirteen and one-half feet in diameter and five feet high. All woodwork was eliminated, mass production in Cape Colony plants was possible, and cost and labor were drastically reduced. Working parties of twelve men and twenty natives could erect an entire shelter and its attendant defensive works in eight hours.

The typical blockhouse line consisted of these shelters spaced at intervals of one-half to three-quarters of a mile and connected by barbed-wire fences. These fences, strengthened with aprons and thick, unannealed steel wire difficult to sever, were anchored in clusters of buried stones. Each blockhouse, costing from £16 (Royal Engineers) to £60 (contract) had several firing loopholes, a door protected by a bullet-proof shield, and a week's supply of water, food, and ammunition. The door led to a sentry trench dug around the blockhouse; ten yards further out was a high wire entanglement completely encircling the shelter. Along the zigzag fences between the blockhouses, wide trenches were sometimes dug to stop vehicular traffic in the event the Boers broke through the fence. Garrisons, connected by telephone, consisted of a noncommissioned officer and five men. A subaltern with ten additional men controlled three blockhouses, and for about every ten houses a captain, with a reinforcing garrison of thirty men, exercised supervision. During the day the connecting fence was patrolled by the blockhouse garrisons, and at night the defenders, warned by flares and empty ration cans fastened to the fences, opened fire on any moving object. When the line was along the railroad, armored trains also patrolled and brought up reinforcements as required; blockhouses on the lines across the open *veldt* were reinforced by mounted units.

In June, 1901, with the railroads now reasonably secure, Kitchener took the momentous decision to construct lines across the *veldt* between railroads. He was aiming at nothing less than subdividing the entire Free State, Transvaal, and the northern portion of Cape Colony into increasingly smaller segments vulnerable to sweeps by mounted troops. By the time the war ended he had built about 9,000 blockhouses covering lines 5,000 miles in length. One of these lines, crossing the desert between Victoria Road Station and Lambert's Bay, was 300 miles long.[46] The important point, however, is that Kitchener did not utilize the blockhouse lines solely as a passive defensive measure. They became an integral part of the drive technique, as has already been noted. They were also an important source of intelligence, helping to pinpoint the commandos as they tried to move about the countryside. And finally, by penetrating into hitherto almost inaccessible portions of the *veldt*, they provided secure lines of communication to important new bases for British columns. It is true that the blockhouses required considerable manpower for construction, but natives provided much of this; the rest of the manpower and the eventual garrisons came from infantry, whose general uselessness in drives against the Boers had already been demonstrated. There was also the problem of the monotony of life in isolation among the garrisons, whose members tended to deteriorate physically and grow edgy and short-tempered. The most significant result of the blockhouse lines, however, was that they contributed to Kitchener's concept of centralization of control and helped give a "mechanical and statistical character to the war."[47] The spirit of aggressiveness so desirable in troops was impaired.

How effective the blockhouse lines were is the ultimate test of their validity. They slowly strangled the Boers, forcing them to split up into smaller parties without wagons, for it was extremely difficult to cross the lines with wagons. Small parties, driven to operating at night, could break through a line, but they usually took more casualties in doing so than they could afford. The blockhouses were much like the spider's net slowly woven around the fly and ultimately entangling him. DeWet, for lack of an appreciation of what the concept of the lines meant, might scoff at the inability of the lines to corral him; but other Boer leaders testified to their efficacy in slowly throttling the guerrilla movement.[48] For another British military measure, however, DeWet expressed more respect.[49]

Night Operations

The night raids carried out by a few of the British column commanders were an effective adjunct to Kitchener's methodical drives. These raids, "escaping [Kitchener's] iron direction in every detail [and] . . . employing a dash, skill, and hardihood which compared with that [of the Boers]," struck at guerrilla camps at dawn.[50] The system depended upon "hands-upper" scouts and agents in Kaffir *kraals* and utilized a sophisticated intelligence network to gather information. A mounted party of about 1,500 men would ride out of its bases at night (according to a secret plan based upon the latest information) and march to a designated Boer camp, timing its arrival so that a surprise attack could be launched at dawn. The Boer was at a disadvantage when caught unexpectedly out of saddle, and he detested these raids. He seldom attempted to defend his *laager* longer than necessary to gain his mount and escape. Furthermore, he tired of the harassing tactics, became apprehensive, and could not afford the nibbling attrition of the raids. None of the raids were decisive in capturing most of the Boers in any given camp, and, inevitably, the insurgents resorted to deception to counter the method. But the raids, in conjunction with the blockhouse lines, contributed to an ever-growing feeling of despondency and hopelessness among the burghers—particularly in the eastern Transvaal.[51]

Perhaps Kitchener would have made greater use of the raid had the war continued, in which case it might have become a more decisive weapon in his arsenal. The intricate intelligence system, the requirement for highly trained troops, and the necessity for imaginative and aggressive leaders, however, would have precluded any attempt at mass application. In any event, the test was never made, for the combination of Kitchener's methods wore down the Boers to the point where they elected to seek peace. Eighteen months after the war presumably had ended, a costly counterinsurgency campaign had been brought to a successful conclusion.

Boer Leaders Accept Defeat

In February, 1902, Kitchener again asked the Boer leaders to discuss terms of a cease-fire. Burgher of Transvaal asked Kitchener for safe-conduct to meet with Steyn. Acceptance of this request brought about an unusual situation whereby the British forces cooperated with Burgher to bring his leaders together while still conducting military operations against the guerrillas. Despite this "summit" meeting, Boer leaders felt that they needed to consult with the military commanders. Consequently, a meeting was held on 15 May at Vereeniging, attended by thirty delegates each from the Orange Free State and the Transvaal. To come together, these representatives from the vast reaches of South Africa, constantly crossed British lines with white flags and safe-conduct passes.[52] After heated debate, the group realized that defeat was inevitable, and they capitulated, influenced by Kitchener's private assurance that he would try to secure the most lenient terms possible.

The final articles of surrender required the Boers to acknowledge British sovereignty. Further, the British stipulated that (1) no burghers would be prosecuted except for war crimes, (2) the Dutch language could be taught in schools and used in courts of law, (3) the people could keep rifles for personal protection, and (4) self-government would be permitted "as soon as circumstances permit it." Great Britain further promised to pay £3 million in aid to the damaged states and to make loans to Boer farmers.[53]

Conclusions

The foregoing analysis of Kitchener's methods permits several generalizations regarding the British solution to the problem of how to defeat the Boers. First, to combat the tactics of guerrillas, the methods and organization for purely formal warfare alone were not adequate. Simply to survive, the irregular had only to exploit his strong points and refuse pitched battle unless he had an inordinate advantage. Against the Boers, countering this situation meant emphasizing mobility, in which the British were initially woefully deficient.

Second, a nation that becomes involved in a counterinsurgency effort should carefully take the long view in plotting its strategy. There are several facets to this generalization. Public opinion at home must be educated to the fact that the action will be frustrating and drawn out, particularly if the guerrillas have a stirring cause and resolute leadership. Both of these the Boers possessed. Kitchener's requests for reinforcements and his harsh measures against the local population were received with some shock in England because of a failure in this area. Similarly, an inflexible stand on terms of unconditional surrender may prove incapable of implementation. In the Boer War, the position taken by Milner—with the government concurring—conflicted with Kitchener's personal ideas and led to bad feeling.[54] This raises the additional point of direction of national effort. In a guerrilla struggle, parts of the country will usually be semi-pacified and under civilian control. The policies applied and the degree of control the military command exercises therein must be clearly delineated. Such an arrangement existed only haphazardly in South Africa. Kitchener wanted to win the war, and was accused of ignoring the political aspects. Milner objected to some of Kitchener's intimidation measures and pleaded for a reconstruction effort free of military interference; he wanted this effort to commence well before the final peace. How much force, and how to apply this force, are correlative problems. Kitchener wanted free rein. There was no one on the spot to coordinate the conflicting views.

From the viewpoint of the purely military measures adopted by the British, Kitchener's methods brought results—but at a price. As a first consideration, he was spared the embarrassment of outside help reaching the Boers. Consequently, he concentrated upon attacking local sources of supply. The success of this effort has been noted; but the concomitant policy of evacuation of dependents may have left a scar on the Boers that contributed to a postwar anti-British feeling of long duration. This is difficult to ascertain positively because the Boers had definite feelings of this type before the war; how much more fuel was added to the fire by Kitchener's methods is debatable. Second, the policy of intimidation, particularly when based upon bluff, did not produce any significant results. If anything, it inspired more defiance. Threats are meaningful only if capable of implementation. Third, Kitchener's policy of centralized control and methodical war discouraged the very initiative, dash, and original thinking needed to defeat the Boer at the column level. In spite of it, however, the war was won; and considering that the guerrilla phase lasted only a year and a half, perhaps no other method would have been more efficient. Kitchener was not blessed with the special type of mobile unit needed to counter the Boer when he assumed command; to train and develop such units and then release them in killer-hunts would have taken many months. It is questionable that he had this much time. He elected instead to bring to bear technology and the might of the British Empire to exhaust his enemy, all the while expressing a willingness to negotiate within the limitations imposed upon him. And he succeeded, as President Schalk Burgher of the Transvaal dejectedly noted at war's end: "We had a confidence in our own weapons; we underestimated the enemy; the fighting spirit had seized upon our people; and the thought of victory had vanquished that of the possibility of defeat."[55]

Notes

1. Rayne Kruger, *Good-Bye Dolly Gray: The Story of the Boer War* (Philadelphia: Lippincott, 1960), p. 365.
2. R.C.K. Ensor, *England, 1870–1914* (New York: Oxford University Press, 1963), pp. 226–232, 245–251; Eric A. Walker, *A History of Southern Africa* (London: Longmans, 1967), pp. 454–486; Godfrey H.L. LeMay, *British Supremacy in South Africa, 1899–1907* (New York: Oxford University Press, 1965), pp. 26–34; Kruger, *Dolly Gray*, pp. 50-55.
3. LeMay, *South Africa*, p. 33. Quotation from Colonial Office papers.
4. Erskine Childers, *The Times History of the War in South Africa, 1900–1902* (London: Sampson Low, Marston, 1907), V: p. 191; Kruger, *Dolly Gray*, p. 470; Christiaan R. De Wet, *Three Years War* (New York: Charles Scribner's Sons, 1902), pp. 313–315.
5. Childers, *Times History*, V: 589.
6. Kruger, *Dolly Gray*, pp. 50, 369, 388; LeMay, *South Africa*, pp. 31–32, 35–37; Walker, *Southern Africa*, pp. 496–497.
7. W.K. Hancock, *Smuts: The Sanguine Years, 1870-1919* (New York: Cambridge University Press, 1962), pp. 111–112; Ensor, *England, 1870–1914*, p. 252.
8. Childers, *Times History*, V: 45, 67. The Boers at this time could call upon 60,000 fighting men, most of them armed and mounted; however, as time went on there were seldom more than one quarter of this number on a war footing simultaneously. Opposed to this number, Kitchener at first had about 210,000 troops, only 10 percent of them mounted; about half of them were scattered in isolated garrisons or along the railroads as guards.
9. U.S. War Department, A.G.O., *Reports on Military Operations in South Africa and China* (Washington: Government Printing Office, 1901), p. 47.
10. L.S. Amery, *The Times History of the War in South Africa, 1899–1900*, 5 vols. (London: Sampson Low, Marston, 1902), II: 66–70.

11. Childers, *Times History*, V: 82; Deneys Reitz, *Commando: A Boer Journal of the Boer War,* 2nd ed. (London: Faber and Faber, 1950), p. 187.

12. The details of the Boer military system outlined in this and the two succeeding paragraphs are based upon: Amery, *Times History*, II: 48–96, and Howard C. Hillegas, *The Boers in War* . . . (New York: D. Appleton, 1900), chaps. IV, V.

13. Ben Viljoen, *My Reminiscences of the Anglo-Boer War* (St. Louis: Becktold, 1905), pp. 349–350; De Wet, *War*, p. 57. See Roland W. Schikkerling, *Commando Courageous* (Capetown: H. Keartland, 1964), p. 379; Frederick H. Howland, *The Chase of De Wet* . . . (Providence: Preston and Rounds, 1901), p. 190, and Kruger, *Dolly Gray*, pp. 243, 249, 338, for additional examples of the independent attitude of the Boer.

14. *Report of His Majesty's Commissioners Appointed to Inquire . . . the War in South Africa* (London: His Majesty's Stationery Office, 1903), II: 171.

15. Amery, *Times History*, II: 48.

16. U.S. War Department, *Reports*, p. 243.

17. See, for example, Childers, *Times History*, V: 532–534, for an account of a mass charge by some of De la Rey's men.

18. Amery, *Times History*, II: 63.

19. *Report of His Majesty's Commissioners,* I: 220, 294; II: 319.

20. Hillegas, *The Boers*, p. 65.

21. Hillegas, *The Boers*, pp. 105–115; Howland, *De Wet*, p. 179.

22. Childers, *Times History*, V: 2–4; R.M. Holden, "The Blockhouse System in South Africa," *Journal of the Royal United Service Institution*, XLVI (January–June 1902), 483.

23. Childers, *Times History*, V: 5; quotation from L. March Phillipps, *With Rimington* (London: Arnold, 1901), p. 123.

24. *Parliamentary Papers*, Cd. 902. (December 1901), quoted in A.C. Martin, *The Concentration Camps, 1900–1902* (Capetown: H. Timmins, 1957), p. 3.

25. Viljoen, *Reminiscences*, pp. 334-335; Childers, *Times History*, V: 92–93.

26. Childers, *Times History*, V: 76–77, 86–88.

27. Philip Magnus, *Kitchener: Portrait of an Imperialist* (New York: Dutton, 1959), p. 179; Martin, *Camps*, pp. 5–8.

28. De Wet, *War*, pp. 409, 415. Details of the outcry of protest over the concentration camps are beyond the scope of this paper. Documented, though not unbiased, accounts appear in Martin, *Camps*, and Le May, *South Africa*.

29. Reitz, *Commando*, pp. 149–150; Hancock, *Smuts*, p. 120; Le May, *South Africa*, p. 89.

30. John F.C. Fuller, *The Last of the Gentlemen's Wars* (London: Faber and Faber, 1937), pp. 171–172.

31. *Report of His Majesty's Commissioners*, II: 319.

32. Kruger, *Dolly Gray*, p. 402; Childers, *Times History*, V: 88, 162.

33. Kruger, *Dolly Gray*, pp. 472-473; Martin, *Camps*, pp. 14-15; Childers, *Times History*, V: 586–590.

34. See De Wet, *War*, pp. 406, 413, for references by Boer Leaders at the council discussion preceding the peace that expresses concern for dependents in the British camps.

35. In January 1901, The Boer leaders met and evolved a strategy to invade the two crown colonies, ultimately aiming at stirring up rebellion therein and bringing the two colonies into the war against the British. Two Boer columns had entered Cape Colony the previous month, and De Wet tried twice in 1901 to reinforce them. That same year Botha moved into Natal for a brief time. When the war ended Smuts had a force in the Cape Colony and was trying to kindle a revolt. See Childers, *Times History*, V: 28– 43, 126, 132–137, 318–320, 333–334.

36. Le May, *South Africa*, pp. 100, 114–121; Kruger, *Dolly Gray*, pp. 414, 429, 466; Hancock, *Smuts*, pp. 133–135.

37. De Wet, *War*, p. 250; Le May, *South Africa*, p.104; Childers, *Times History*, V: 322.

38. For details of the drives see Childers, *Times History*, V: 131–157, 158–182, and Kruger, *Dolly Gray*, pp. 380–387.

39. Childers, *Times History*, V: 247.

40. *Report of His Majesty's Commissioners*, II: 27–28.

41. J. Watkins Yardley, *With the Iniskilling Dragoons* (London: Longmans, 1904), pp. 342–343.

42. Yardley, *Dragoons*, p. 305; Childers, *Times History*, V: 468.

43. Yardley, *Dragoons*, p. 320; Childers, *Times History*, V: 469–472.

44. Childers, *Times History*, V: 473– 491.
45. The factual aspects of the description of blockhouse lines in this and the three succeeding paragraphs are based upon: Holden, "Blockhouse System"; Childers, *Times History*, V: 256–260, 396–403; Fuller, *Gentlemen's Wars*, pp. 107–111; *United Service Gazette*, No. 3603 (1 February 1902), 88–89.
46. Fuller, *Gentlemen's Wars*, p. 107.
47. Childers, *Times History*, V: 263–264.
48. De Wet, *War*, pp. 260–263. Comments of other Boer leaders appear in the account of the Boer council on the eve of peace (See *ibid.*, pp. 335–346, 355); also see Viljoen, *Reminiscences*, pp. 311, 383–389.
49. De Wet, *War*, p. 263.
50. Kruger, *Dolly Gray*, p. 416.
51. Kruger, *Dolly Gray*, pp. 440– 441; Childers, *Times History*, V: 451– 452, 459.
52. Martin Blumenson, "South Africa (1899–1902)," D.M. Condit and Bert H. Cooper, eds., *Challenge and Response in Internal Conflict* (Washington: American University Center for Research in Social Systems, 1968), III: 77.
53. *Ibid.*, p. 78.
54. Le May, *South Africa*, pp. 121–122.
55. Kruger, *Dolly Gray*, p. 477.

══════════════════════ COMMENTS ══════════════════════

As we approach the close of this anthology, we are reminded again that it is fitting for the military historian to be modest about the uses of his discipline. The British Army found the Boer War distasteful enough that it did not study the war as much as it might have; but in the generation now just ending, the British Army has had so many experiences in waging counterinsurgency war against opponents employing unconventional methods that it has accumulated a vast store of historical knowledge at first hand. If any army has been steeped in the military history of waging counterinsurgency war, it is the British Army of our own day. And as J. Bowyer Bell indicates in his 1973 presentation printed here, the British Army has deliberately, thoughtfully, self-consciously sought to digest and ponder upon its recent historical experience. As Dr. Bell also indicates, however, a wealth of relevant historical knowledge has done no more than permit the British Army to cover its successive withdrawals with a certain ritualistic grace. The Army's experience has at best merely permitted it to slow the pace of the passing of empire to that of a stately minuet.

Dr. Bell is affiliated with the Institute of War and Peace Studies of Columbia University. He confesses to an infatuation with insurrectionists and has written about them in many geographical settings; but he also confesses that he prefers doing research about unbenign events in benign climates such as Ireland's, so he has written especially about that country, as in **The Secret Army: The IRA, 1916–1970** *(New York: John Day, 1971).*

Revolts Against the Crown: The British Response to Imperial Insurgency

J. BOWYER BELL

In JANUARY, 1944, an illicit proclamation began circulating in the British Mandate of Palestine. This declaration of a Jewish revolt by the Irgun Zvai Leumi was to be a harbinger of a generation of imperial insurrection, the first lines written in the final act of the Empire. In 1944 an "armed struggle" by a small "military" arm of the schismatic Zionist Revisionist Movement neither impressed nor particularly concerned the British, nor did the first marginal operations. A vast world war was under way, in part directed against the Zionists' most dedicated enemy, Adolf Hitler. All the orthodox Zionists opposed the antics of the little group of zealots in the Mandate—a minority of a minority, who frightened no one. What to the British was surprising was that an open revolt, however ineffectual, had been launched by these fanatical Jews against their old "ally." The British were shocked, surprised, outraged, and indignant at the pretensions of a gang of terrorists who were without legitimacy or popular support.

But no matter what the British moral response, the tiny revolt escalated year by year into a massive emergency that drew in tens of thousands of troops, ate up precious sterling balances, alienated old friends, even the Arabs, and ultimately engendered profound disgust on the part of the British public. In pique and desperation the British sought recourse in the United

Nations, and in 1948 in general disarray evacuated the Mandate. By then the armed struggle of the Irgun Zvai Leumi had become a classic. Under Menachem Begin, the Irgun had devised a strategy that became a model for imperial revolt. A means had been discovered for the weak to lever out the strong—or so it seemed to some.

Before 1944, the British Empire had been exposed to two serious experiences of national revolt. In America in 1776, the rebels, benefiting by distance and major allies, created alternative institutions and defended them by conventional means. Learning in part their lesson, the British during the nineteenth century slowly evolved a counterstrategy to rebellion by those sufficiently mature for self-government: the devolution of power by central concession to new dominions. This in the twentieth century had been refined into the Commonwealth strategy, immediately effective in the English-speaking Dominions and potentially applicable elsewhere. By then an alternative rebel strategy had been devised by the Irish, who in one guise or another, by the application of an entire spectrum of techniques and tactics, had for hundreds of years been engaged in an effort to create an Irish Ireland. The Irish, rather than the distant American, experience became a primer for potential rebels elsewhere in the Empire who did not appear to be designated candidates for the Commonwealth strategy.

In 1916, in what had then seemed the last gasp of the militant Irish Republican Movement, a creed that repeated British concessions to the Irish had nearly undermined, a traditional Rising wracked Dublin during Easter Week. This Easter Rising, as had all others, collapsed into bitterness and recrimination—apparently the gun had at last left Irish politics. Beginning in 1918, another more thoughtful attempt was launched by a younger generation. They attempted to create an Irish Republic, Free and Gaelic, by coupling irregular war led by an underground Irish Republican Army (IRA) with the creation of alternative governmental institutions. The subsequent British repression could prohibit the Republican institution from functioning but could not crush the IRA "terrorists"—in fact the increasingly stringent measures taken against the IRA became distasteful to the British public. In time the British found a means of compromising the issues with a formula that created an Irish Free State in twenty-six counties, a loyal, largely Protestant enclave in six counties of Ulster, and a guarantee of British bases and economic interests. Eventually the Free State evolved into an Irish Republic outside the Commonwealth; but in Britain the Irish Treaty was viewed as a splendid exercise, if unique, in the accommodation of national aspirations, a judicious application of a strategy of devolution. The Irish strategy of revolt, if not emulated in the other parts of the Empire, more distant but less bold, less organized, was not forgotten. Few potential nationalist rebels could hope for the assets of a George Washington, but all could in the pinch manage murder from a ditch.

By and large, however, it was not the cunning Irish combination of terror, British hypocrisy, shadow institutions, international propaganda, guerrillas in the hills, the exhausted imperial machine, and the war-weary population that potential rebels studied. The real key to national liberation appeared to be in India, where Mahatma Gandhi and Jawaharlal Nehru had fashioned a mass movement based on a disciplined, nonviolent campaign of civil disobedience. The leaders of the Indian Congress Movement were convinced that once the masses were motivated, disciplined, and determined, the British faced by general civil disobedience would have no choice but to rule by the most brutal and self-destructive force or concede. And they suspected that the force to coerce 400,000,000 people did not exist, and even if it had, the will to employ it did not. In 1935 the British had admitted as much with the passage of the India Act, and in 1942 in the midst of the war, London had in effect promised independence to the Congress leaders. For most nationalists the Indian strategy seemed to offer the most, for it neatly fit into the British Commonwealth strategy, peacefully demanding what should be cheerfully granted, avoided the risks of open revolt by the weak, and offered a means to mold the future nation through disciplined political activities.

At the end of the Second World War, then, within the British Empire there were two major nationalist strategies: that of leverage, based on an armed struggle of attrition, the Irish-Irgun option; and that of India, dependent on civil disobedience on a vast scale by a disciplined mass party. And there was the British alternative in the Commonwealth strategy, a means of devolution of power to the mature. At the time, the strategies of the orthodox revolutionaries, the Marxist-Leninist rush by the urban proletariat to the barricades, and the distant experience of Mao Tse-tung in rural China, appeared alien to imperial experience. In the course of the next generation, the British Empire would disappear—a massive act of devolution that passed largely peacefully. The Indian strategy was applied in all sorts of odd corners and with varying exceptions became the conventional means to power, however much the process might have been accelerated by open revolt elsewhere. After the Indian success, it very soon became clear that even in less mature colonies like the Gold Coast progress was possible. There Kwame Nkrumah effected the independence of Ghana by adopting similar methods, less disciplined, more disruptive, but in time equally valid.

There were exceptions. For varying reasons the process of devolution did not always run smoothly. The Malayan Communist Party (MCP) launched a guerrilla war using the strategy of Mao Tse-tung and the enthusiasm and ambitions of the local Chinese community. In Kenya the Kikuyu, outraged by colonial policies and the "theft" of their land, attempted to combine the politics of agitation led by the Kenya African Union (KAU) with the terror of tribal violence loosely organized as the Mau Mau. In Egypt the various political factions sponsored fedayeen raids into the Canal Zone to coerce British concessions. In Cyprus, Colonel George Grivas organized a resistance movement, EOKA, and in collaboration with Archbishop Makarios sought unsuccessfully to achieve union with Greece, *Enosis*. In South

Arabia the militant Arab nationalists, emboldened by the direction of events after 1956, launched an armed struggle certain that Britain's moment in the Middle East had passed. And in 1967 the triumphant National Liberation Front (NLF) established after the British departure the new People's Republic of South Yemen. There were as well other rebellions and disorders and continued imperial responsibilities that found British troops active in Borneo or Oman and even between 1956 and 1962 in Ulster against the IRA; but the Gold Coast aside, the major imperial emergencies faced by the British were Malaya, Kenya, Cyprus, South Arabia, and in a special way Egypt. All were very different indeed—the MCP had recourse to the strategy of Mao, and the Mau Mau to atavistic tribal custom—and yet the British response became a pattern sufficiently predictable to be ignored at rebel risk, a pattern so firm that even after the end of imperial insurrections the response in Northern Ireland in the seventies to renewed violence by the IRA appeared to come from the same imperial mold.

That the Commonwealth strategy did not work everywhere was mainly, the British assumed, either because the imperial power was responsible for adjusting conflicting claims, those of the Arabs and the Jews or the Greeks and the Turks, or for eliminating unrepresentative claimants, the Communists in Malaya or the Mau Mau in Kenya. In any case those gunmen and terrorists who sought power outside the Commonwealth route were without legitimacy. Thus for the British a revolt opens really not with bombs but with the unexpected surfacing of a conflict over legitimacy, a conflict that in most cases has had a long and troubled history, a history cherished by the rebel and ignored or denied by the British.

The immediate British reaction to the rebels' aspirations, no matter what the circumstances, is outraged indignation. The rebel is an alien and evil man, motivated by personal ambitions, often deluded by an imported ideology, who uses terror to acquire support—a man outside the law, outside common decency, outside reason. The full majesty of historically recognized, internationally accepted, *legitimate* authority is turned on the little band of assassins. In the long run rebel legitimacy can only be won by force—or by concession. Some of these illegitimate claimants could, as had been the case with the Irish in 1921, be co-opted by means of an adjusted Commonwealth strategy; but in some cases there was nothing for it but repression. After 1944 in some foreign corner or another, regularly to their surprise, the British had to pursue an anti-insurgency campaign led by indigestible rebels. British colonial officials, career officers, and policemen might, if they were keen, usually serve in several emergencies. Some appear a little grayer, a little wiser, a little further up the ladder, in each new campaign, like spirits of revolts past. The British knew and continued to have their knowledge reinforced of the dangers and costs of such revolts and the means to avoid the worst problems. From their exposure the British learned the tactics of anti-insurgency, the cost of an emergency, the importance of political concessions, and the means to manipulate the Commonwealth strategy. The Cabinet of whatever composition knew the cost of staying or getting out. Still the British, caught every time by surprise when a revolt did begin, were every time shocked, outraged, indignant.

Only rarely did the authorities, either on the spot or in London, foresee the possibility of an armed revolt. Conditions that the potential rebels felt were intolerable, that created deep frustration, and that could not be ameliorated except through violence, did not so appear to the British. In many cases the British often could not conceive of priorities different from their own.

In Palestine the British simply did not understand the impact of the European Jewish Holocaust and the depth of Jewish agony, nor could they credit the charge of collusion with genocide made against them. In the Gold Coast the motive of the mob—political power—went far beyond the usual bread-and-butter issues of colonial politics. The British had simply not dreamed that such factors would appear in the colony for decades. In Malaya the revolt by the MCP was launched not from the depths of despair as in Palestine, but from the high ground of ideological certainty—native Chinese ambition in Malaya hued over with a vision of a Communist future. In this case the British were surprised that the MCP dared to revolt rather than at their aspirations. In Kenya the European settlers and local observers had *feared* a revolt but had not anticipated one. Thus while following policies after mid-1952 that almost insured a Kikuyu "revolt" would take place, there was still surprise at the extent of Kikuyu alienation in Nairobi, while in London the new emergency had been quite unanticipated. In Egypt no one had feared a revolt and few had been surprised at the fedayeen forays, only slightly outside traditional Egyptian sideshows; nor was the eventual accommodation, another treaty, novel. What would come as a bitter surprise was the Suez humiliation, but that fell largely outside colonial considerations. Long after the Cypriot emergency was over, British spokesmen of various hues insisted that *Enosis* was and always had been an artificial issue exploited by agitators. By so refusing even to consider the matter, the British, knowing what the Greeks *really* wanted, had set a boundary to nationalism that someone, sooner or later, would cross—as Grivas did to British surprise. By the time of the South Arabian misadventures, Britain should have been beyond surprises at the ambitions of radical Arabs; but even though Radio Cairo reached into the hills of Dhala, the British still hoped that the old ways and old forms would work with the new Arabs and were surprised and indignant when they did not.

The British difficulty in perception was a fault hardly limited to the British, since surprise had long played a commanding role in military and political affairs. In some cases the potential rebel intended to take up arms no matter what accommodation was offered; but there at least the British might have been forearmed. Even if the rebels did in fact represent alien strains within the Empire, there remains the possibility that a more perceptive eye would have uncovered the pattern of frustration and suggested an alternative to repression. It is, to be fair, difficult to see how London heeding Cassandra could or would have acted much differently in most cases. The rebels largely felt impelled to revolt because a nonviolent dialogue no longer offered them anything. In Palestine the whole direction of British Middle East policies since 1939 largely precluded undue concessions to the Zionists, and for the men of the Irgun no concession, however generous, would have done. In Malaya the MCP's conviction that victory was certain, and any course but the armed struggle dangerous, would probably have remained no matter what the British did or did not do. In Kenya, at least, a realization of the nature of the most immediate Kikuyu grievances might have allowed time for a Gold Coast dialogue to evolve; then again, given the settlers' attitudes, perhaps the necessary concessions were out of the question at the time. In Cyprus the British might have taken *Enosis* seriously, sufficiently so in any case to point out the international complications that might ensue and the rigid requirements of British security—but would the patriots have listened? And surely no concession would have swerved the Arabs in Cario and Aden from their allotted course. Almost nowhere then, except perhaps in Kenya, could a dash of prescience have greatly altered the situation; for the rebels wanted to rise in arms for purposes quite beyond the capacity of the British to concede.

In most cases the closer the individual was to the scene of the action on the eve of the trouble, the more likely the chance of error, the failure to perceive change. Those who knew the most often saw the least. The-Man-Who-Knew-The-Natives often missed the impact of modernization or the influence of new ideas. Often he had learned his job and about the natives on the spot, acquiring the rare and esoteric languages of the bush, absorbing detailed and extensive anthropological data, fashioning a career on extended tours. Some of the "natives" in Tel Aviv or Nicosia were quite different from those of previous colonial experience. Elsewhere the attractions of education, the appeal of Western technology wrought swift changes, stirred quite "unnative" ambitions, tilted the familiar into new and not always visible patterns without ever showing the British on the spot a new face. And when the face did appear above ill-fitting white collar and obscure school tie, few realized just how profound the change and how limited the old means of control. Even when that control crumbled, there was only limited understanding of what had gone wrong, that the natives had given up the effort to take part in a dialogue with the deaf and sought recourse with bombs.

In carrying on the imperial dialogue in many places, the British had been talking without listening, looking at events without seeing. As the years passed, the discontent turned to the more lethal dialogue, a strategy that inevitably came as a surprise to the British. Mass nonviolence, the politics of confrontation, the tactics of direct action, first in Asia and then in Africa, did not only surprise the British but also caught their attention. The British monologue died down, and the new native voices of the Gold Coast or Egypt could be heard. If the means of interrupting the British monologue appeared illegitimate, Mau Mau oathings or Grivas' bombs, or if the time to attract British attention was too short, Palestine in 1944, or if the rebels did not care to talk, which was mostly the case, then Britain would be surprised at the new form of communication—a revolt by the natives no one ever knew.

After surprise at the new lethal dialogue came shock that rebels would seek recourse to violence when means of accommodation abounded, when the expressed grievances were not legitimate, when the mass of decent people disapproved. Without exception the first analysis on the spot and then in London was that the revolt was the work of a tiny disgruntled minority, dependent on support achieved by coercion or intimidation or violence.

This British analysis was almost always in part correct; for revolts, certainly at the beginning, are the work of a tiny handful of men acting in the name of the masses who, of course, can hardly be polled. The Irgun, EOKA, the Egyptian fedayeen, the NLF in gross numbers were tiny and remained so until the end. The British approach was that because the revolutionary organization was small, it was also unrepresentative. And this, too, was often true. The emergency in Kenya was as much a Kikuyu civil war as an armed insurrection, and it hardly involved most of the other Kenya Africans. The Irgun were a self-confessed minority in a Jewish community that in turn was a minority in the Palestine Mandate. The Communists, in the jungles of Malaya, tied in a support community of less than a majority of the Chinese, who were again a minority in the colony. Most revolutionary movements, where reasonable estimate is possible, always have been led by a tiny minority and actively, even passively, supported by less than a substantial majority. Often the rebels must coerce or eliminate the loyalists. In South Arabia, for example, more Arabs were killed by the rebels than by British security forces. Thus the British were quite right: the rebels were a minority with limited active support and probably limited support of any kind. The British too, if reluctantly, noted that some support must exist, for information and intelligence about the rebels proved difficult to acquire, and public expression of gratitude for British counterinsurgency efforts was limited.

The obvious conclusion was that the minority was intimidating the majority. That rebel support must be the result of intimidation does not, however, logically follow. Many Jews in Palestine, for example, did not approve of the Irgun's campaign but would not oppose it. Many did not want to be either informer or advocate. Much the same was the case with the Greek Cypriots, of whom many who preferred the quiet life would not oppose EOKA. In fact, much if not all the mass always seems to tend toward the quiet life. Given a chance, they would vote for a truce or a pause; given no chance they permit the rebels to sacrifice for a higher national "purpose" beyond the ballot box. This neutrality, a slightly biased neutrality, however, is all that a rebel needs. A government needs more; for if the rebel continues to exist, he will in time win—while a government must govern, must win outright, must restore order and hence law. The British problem was to woo the vast apolitical audience. This audience is often only marginally interested. This passivity is a British loss and a rebel gain. Not only has the dialogue been broken by violence, but the medium of order, the acquiescence of mass, becomes suspect, and rightly so, in British eyes. The rebels paradoxically need not organize the masses, but the British must to deny the ocean to the fish.

The British assumed that only through force could the rebels achieve toleration. This often was the case: neither EOKA nor the MCP nor the Kikuyu made much pretense that they were not executing traitors and informers; rather the reverse, and they spread the word. Thus the British assumed that at heart the population supported them, not the rebels. This attitude is not British counterpropaganda but an article of faith. The British believed and so acted. Trust is maintained in the "real" people, and outrageous risks are taken because of this trust. In Nicosia the valet slipping a bomb under the bed of Field Marshal Sir John Harding, the Military Governor, was by no means a unique betrayer of that trust; there were repeated betrayals. Everywhere from Malaya to Aden, the potentially disloyal servants were kept, often to the last day and the ultimate betrayal. Even the frantic settlers in Kenya wanted "to kill every Kikuyu but ours." Since British authority, then, is legitimate, all good men and true will rally about in opposition to the illicit pretenders—unless so prevented by violence. And the rebels are violent, illegitimate men who at best can count on the dubious virtue of certification from a recognized revolutionary center in Cairo or Moscow or from a greedy regime in Athens.

The British had to stand for something as well as against sin. The simple legitimacy of being in previous possession of power is insufficient once the old dialogue has broken down and the violence begun. To undergird the banners of imperialism and the primacy of the Empire, the British stood for order, decency, fair play, good government, civilization, justice, and law, and occasionally Christianity if appropriate. And, of course, this was true; they did. What Britain did not stand for was immediate independence and native interests over those of Britain. This was the rebel program—everything for us now—and it had great charm. For the British to insist that immediate independence would be disastrous and that Britain could do more for the natives than they could do for themselves would not go over well. Whatever Britain's position on self-determination, now or later or never, the rebels had to be shown as men who would use proud slogans for low purpose or wave the national banner while selling the nation abroad to alien ideologies. The rebels are thus not nationalists but illegitimate pretenders to power that they intend to misuse—men past the stage of foolishness into knavery, and criminal knavery at that.

If the revolt is absolutely illegitimate, totally without moral justification, led by men without scruple or decency, then it is obviously both easier to oppose naked evil and also quite difficult to withdraw, almost impossible to compromise. In all conscience, it is difficult to take tea with a terrorist or accept a criminal into the palace; much more important, however, it is far easier to seek out and destroy evil. From the first, the leadership of the revolt is defined not simply as evil but as alien. And the more appropriate the label the more likely that the establishment of order will be pursued with maximum force. And the "cause" that led to the open revolt has also been perceived by the British as alien, spurious.

If not exactly spurious, the rebel causes over a generation have certainly been indigestible. While some nationalists needed and accepted the slow process of institution-building within the Empire and accepted the British Westminster model, whether appropriate or not, having no other, the rebels did not. They felt no need for British models and little value in a Commonwealth dialogue in the British example; all their ideologies, Zionism, Communism, or Hellenism, were alien to Empire.

The most alien of all enemies of Empire were the Mau Mau, an atavistic descent into savagery, absolutely illegitimate in political terms. The toll of Mau Mau killed, the mass detentions, resettlements, and imprisonment could all the more easily be undertaken in light of the horror of the oath and the brutality of the massacres. No one seemed particularly surprised that over one thousand Mau Mau were executed in contrast to only eight members of the Irgun during the Palestine emergency. It was, no matter what the provocation, *hard* for the British to kill Jews. Much the same was true in Cyprus, where the British were fond of the Greeks and deeply frustrated that EOKA could not see where their struggle for *Enosis* was leading. The British were not fond of the Chinese Communists in Malaya—an international conspiracy of an alien ethnic group that threatened British security and the future of Malayan development. The MCP was absolutely illegitimate. The Chinese were not mad, as were the Mau Mau, but they had been converted to an alien ideology. So, too, in Malaya there were the detentions and arrests on a vast scale, the deprogramming camps, the wholesale movement of population, and the huge toll of dead terrorists, executed or killed in the jungle sweeps. The more effective the British were in defining the rebel as alien—and if as a rebel you start by being a Kikuyu instead of a Greek the process is simpler—the more likely the authorities were to see a polarized conflict without solution and without a need for excess compassion; then rigorous repression became the order of the day.

Naturally, once this policy of repression begins to show results, a reversal to allow for accommodation becomes difficult, a lengthy procedure of redefining and reconsidering. In the Gold Coast the Watson Commission investigating the disturbances of 1948 produced a report that depicted Nkrumah as imbued with a Communist ideology that only political expedience had blurred. A red knave with a black skin quite obviously would not be the man for the future. There was in the Gold Coast sufficient leeway for Nkrumah to apply the Indian strategy, adapted for African use, and most important there were British officials in Accra and London who recognized what he was doing. In Kenya, even in 1960, Jomo Kenyatta was still the leader of darkness and death to Governor Sir Patrick Renison. In two more years Kenyatta would be a senior minister serving in an African administration presided over by the same governor. The process of turning Kenyatta, a man "definitely guilty" of leadership of a murder cult, into the doyen of the new African statesmen took not only time but also the pressure of expediency—still it was done. Nor if the nationalists in South Arabia had played the game is there much doubt that the Arab terrorists, thugs and pawns of Egypt, could have been transmuted into candidates for the Commonwealth conference. It is not a process without pain, for many persist in seeing yesterday's villain behind today's glory. Through practice and experience, however, the British could reverse the policy of alienation.

At the time of the revolt, if the rebel is alien, the alternatives appear exacting: crush the revolt or evacuate. Both the Mau Mau and the MCP were crushed and remained in British eyes primitive tribesmen and pawns of Communism. In South Arabia the British managed only a semblance of devolution in a last-minute agreement at Geneva; in effect London threw the keys over the wall and evacuated, leaving a long line of Arab friends who had been led up the garden path. In Cyprus the Commonwealth strategy finally came to the rescue. In Egypt a typical treaty solution in 1954 led only to 1956 and the last violent hurrah of Empire. In Palestine, as in South Arabia, the British scuttled, but under the auspices of the United Nations. Still two clear wins, two clear losses, and two revolts accommodated is not a bad show for either the military or the diplomats.

Even when facing what appeared alien, unrepresentative, and illegitimate power-grabs, the British did not rely on coercion alone. Experience had revealed to all that terror could best be countered by political maneuvers. Thus if the rebels were not *too* alien, the Irgun or EOKA or Arab nationalists in Aden, then the British sought a political option to involve the forces concerned. Such a political strategy might exclude compromise with the rebels but might produce an accommodation that the rebels would accept, as was the case with Cyprus. Even in Malaya and Kenya, parallel political programs were launched to erode the support of the rebels, even while such "support" was officially denied. It was, of course, easier to reach an accommodation, no matter what sort, when the perceived level of rebel violence was low. The Mau Mau violence level was perceived as frightful—

which in the number of European casualties it was not—while the fedayeen attacks and the subsequent burning of Cairo were labeled as traditional Egyptian troublemaking and easily discounted at the bargaining table. In almost all cases, alien rebel or no, the British sought a small bit of uneven middle ground even when the aspirations of the rebels endangered crucial British interests or seemed beyond any rational accommodation.

British colonial strategy, then, was to open and maintain a dialogue, the British face of the Indian strategy, unless caught by surprise. Then, if the rebels were too alien to be co-opted, Britain would fight it out. Simultaneously, Britain might grant parallel concessions—self-government—to the loyal natives. The tactics Britain used to balance concession with coercion varied greatly and were often the result of independent initiatives, contingency factors, and contradictory impulses. Britain never had an overall book of colonial tactics. The British, on the military side, after long experience, devised a basic approach to both urban and rural insurrection that properly applied went far to reduce to quite manageable proportions the level of violence. As long as a few men were determined on Liberty or Death, there would be some trouble; but the British experience indicated that such trouble could be narrowly limited if the Cabinet wanted to wait long enough, invest enough in repression, and cooperate in devising parallel political solutions. If not, all the military efforts would abort.

Military tactics in the field, refined over a generation in the hands of officers and men who often had several differing experiences in counterinsurgency, could be learned, could be applied, could in the long run be largely effective; but only if used in a cunningly prescribed political formula. Military operations in a political vacuum only created a more efficient rebel playing up to the challenge. British military tactics were, therefore, of little use unless political conditions were factored into the formula, and the political conditions in each case were quite special. Consequently the ultimate formula adjusted to the various regional unknowns was in each revolt different.

British political tactics in pursuing a strategy of devolution, even in the midst of open revolt, varied vastly and could be quite flexible. Naturally, a basic principle was to isolate and if possible ignore the rebels. This meant keeping out international investigators, ignoring United Nations resolutions, and turning back efforts to broaden the crisis. The British, however, if there was advantage, had not the slightest compunction in switching gears. Until Grivas' bombs went off, Cyprus was an internal matter. Almost immediately after the reverberations of the EOKA bombs, there was suddenly a London Tripartite Conference: Cyprus had become an international matter, or at least to the degree that London recognized Turkey's concern; *i.e.*, Ankara had a veto over Greek ambitions. Palestine, too, in time became an international problem before the United Nations, although the British withdrew rather than effect an international solution not to their own advantage. And when all else failed in South Arabia, London snatched at the United Nations straw in hopes that one more committee might produce results. There were, then, few hard and fast rules in the use of political tactics.

The British continued to devise a variety of approaches to each crisis that might support a return to order: constitutions, commissions, royal visits, aid and development, promises and programs. All were used not so much indiscriminately but as part of the uncertain dialogue. Some of the offers might appear to be positive steps, some might be so, some might lead to further devolution, but all played a vital role in forcing the pace. The Macmillan Plan in Cyprus was a non-starter, but it did save face in Athens, where the regime could claim that anything was better than the new proposal. It made a point that was well and quickly taken by the Greeks, who could then disavow EOKA for good and honorable reasons. So year in and year out, the British did divide and conquer, they also united and ruled, found old ways out of new corners and the reverse. Tactically, the political initiatives in colonial matters were inventive, creative, and often effective. That there were so few revolts, and that those so often led to accommodation is witness to British political acumen. The British did stumble on occasion; but still there was enough of that graceful swiftness of foot, so admired and so feared by the lesser breeds, to give evidence that there was life left yet in Perfidious Albion.

British political responses to colonial insurgency were often complicated by British strategic interests—although seldom by economic factors. London often had to pick up the chits for the strategists, maintain possession of the odd bit of real estate the generals needed, unite the impossible to defend the bounds of the indigestible. Essentially, on this level the military laid down that the cost of maintaining a strategic position was worth the price in colonial turmoil. As long as the Cabinet bought the premise, the ground to maneuver politically was limited. The criticism that this turmoil negated the strategic value of the colony in question long remained an article of faith with many imperial critics. Whether or not Britain needed Aden in 1965, Cyprus in 1957, or the Suez Canal in 1952 is quite a different matter. What is quite clear is that the cost of repression, high or low, does also pay dividends. An occupier with sufficient power can put down resistance or contain insurrection and continue to use any base—as the British proved in Cyprus in 1956. Toting up the real cost is much more subtle, however, than satisfying the generals' demands for a fortress. In any event, as the generation of devolution progressed, the review of value received indicated that often the strategic content was less vital than the generals assumed and the room for political maneuver broader; even outright evacuation was in some cases no longer a strategic disaster. Increasingly, too, those in power felt that British economic interests would be as well or better maintained with an indirect presence. Some more pragmatic capitalists had their doubts in certain countries, but hardly anyone wanted to stand up and be counted as an opponent of devolution solely because British profits might suffer. Thus, by the late sixties neither strategic nor economic interests greatly impeded the rush to dismantle the Empire.

Long before then, just as there was a pattern of political response, so had there been a pattern on the part of the British

Army. The first part of it was an avid desire that the political strategy devised in London be forthright, rigorous, and orchestrated with the local military effort. At times, of course, as already noted, the military's own definition of British strategic needs, for example the whole island of Cyprus on a military base, to a large extent determined the bounds of political action. The military recognized, however, that those bounds were limned in London on broad advice, not as a result of the direction of the General Staff. Beginning with the Palestine experience of drift and scuttle, the British Army recognized and was sensitive to both the political aspects of revolt and to the ultimate power of the Cabinet. Secondly, if possible, the Army wanted the power to pursue the rebels with a rigorous campaign centralized under one command, ideally that of a military man, preferably unrestricted by local authorities. Such ideal conditions seldom existed; either there was existing and competing local authority, or serious limitation on the degree of force to be used, or both. Largely, however, the military attitudes toward the rebels differed in no significant way from those of others involved.

In sum, within the process of devolution the basic British response to revolt, honed by experience, not always properly absorbed, remained largely the same. First came surprise, followed by shock and the processes of defining the rebels as a minority of evil men using force to garner support from the basically loyal people. If the rebels were beyond compromise or had taken up arms openly, the Army pursued an anti-insurgency campaign in sure and certain knowledge that their efforts in the field could only succeed if a political formula were found and supported for the long run by London. The means to fashion such a formula might, however, be limited by the very needs of the military, thereby protracting and complicating the problems of suppression. Seldom were economic factors crucial even when in Malaya or India or the Gulf they were important. In large part the British managed to avoid open revolts and even then devised solutions other than absolute suppression or evacuation. Given the number of "natives" involved, the opportunity for misunderstanding, the international interference from various friends and enemies, and the nature of partisan politics, the dialogue of devolution seldom broke down.

With the evacuation of South Arabia in 1967 and the end of a British presence east of Suez, the end of Empire appeared to have arrived. There was a flurry of ceremonial flag raisings on small islands and in tiny enclaves and a spurt in Commonwealth membership. Little was left under British sovereignty but rocks and reefs. There were marginal and distant responsibilities in Oman or Borneo, but with the end of Empire the patterns of response to revolt seemed to belong to the past. This did not, however, turn out to be the case, for once more the Irish question, this time in a particularly violent form, appeared; and once more, whether or not Ulster was an integral part of the United Kingdom, the British response followed the imperial model.

In August, 1969, the civil rights campaign in Northern Ireland collapsed into sectarian violence. The Protestant Unionists, particularly the more militant, believed that "civil rights" was a code word for a united Ireland controlled by a Papist Dublin regime. The Catholics determined to achieve the same political rights available to others in the United Kingdom, the question of a united Ireland aside, would no longer tolerate the provocative and humiliating ritual insults of the majority. They saw this Unionist majority organized into a Protestant state for a Protestant people, where injustice was institutionalized and enforced by a sectarian para-military police force and the weight of biased law. Once the riots that began in Derry spread to Belfast, the provincial government at Stormont, the result of the 1921 exercise in devolution, could no longer control the situation; and London ordered the British Army in to prevent what appeared to be impending civil war. The British, long uninterested and uninformed about Irish matters, were surprised at the level of violence, the depth of communal hatred, and the intractable and primitive differences between Catholic-Nationalists and Protestant-Unionists. This surprise and subsequent distaste did not, however, lead the Labour government to institute radical political measures to ease what were accepted as legitimate Catholic grievances. Instead, change was urged on a reluctant and suspicious Stormont regime. In the meantime the Army kept the uneasy peace, waiting for a moderation of tension or a parallel political initiative. Northern Ireland appeared beyond moderation and, without a startling political initiative, began to slide toward chaos.

Three factors produced an open guerrilla war in less than two years. First, London continued to dither, fearful that reform might inspire violent Protestant response, hopeful against all the evidence that time would heal, sanguine to the point of lethargy. In June, 1970, a new Conservative government came to power, less concerned with forcing change on Unionist allies and without what had at last been a growing sense of urgency within the Labour Cabinet. There were still no radical measures.

In the meantime the Army saw and was allowed to see its mission as the forceful and rigorous imposition of order. No matter how welcome the military had been to the Catholic community in the first place, there could be no doubt that the Army's posture of neutrality would decay in their eyes. The local "legitimate" authorities at Stormont were loyal, Unionists flaunting the Union Jack, proper people who wanted no change. They trusted their police and their army and their friends in London. The Catholics, on the other hand, wanted change, sought it in direct action in the streets, distrusted or actively opposed the "illegitimate" authority of Stormont and the police. They caused trouble, and often did so under the rebel tricolor instead of the Union Jack. The British Army sweeps and searches, the use of armor and gas, the quick hard response was viewed by an increasing number of the minority Catholics as evidence that the Army had been co-opted by Stormont. The Army, as might have been expected, slipped from the role of the defender into that of the oppressor.

The slide, all but inevitable, was accelerated by a third factor: the maneuvers of the Irish Republican Army. After the

August riots in 1969, a new, more military IRA had been formed, the Provisionals, or Provos, in contrast to the politically minded official IRA. The Provos at first accepted the role of Catholic Defender, collecting arms, organizing on a neighborhood basis, and even cooperating with the British Army to maintain the peace. The leadership of the Provos, a mix of old rebels and new militants, however, intended to transform their role to that of an underground army determined on the expulsion of the British. Thus British tactics, often carefully manipulated by IRA provocation, had by early 1971 alienated much of the Catholic minority. The IRA could then move over to an offensive campaign against the enemy, i.e., the British Army.

British problems were only compounded in August, 1971, with the introduction of internment without trial on the advice of the Stormont regime, who had insisted they knew their Irish natives. The minority then withdrew its consent to be governed, went on rate-and-rent strikes, and backed the Provos. By the end of the year, urban guerrilla war in Ulster had reached such a level that the British Home Secretary, Reginald Maulding, admitted in Belfast that the IRA could not be crushed but only held within tolerable bounds.

Over the next six months, the British Army was sorely pressed. IRA bombs devastated much of downtown Belfast and Derry. The roads in the country were unsafe. There were cross-border incidents, constant sniping in the built-up areas, and a parallel wave of Protestant militancy. The killing of thirteen civilians on Bloody Sunday in Derry in January, 1972 attracted worldwide attention and vastly complicated the Dublin regime's effort to contain the IRA in the South. Ultimately, the Stormont Parliament, utterly discredited, was prorogued and direct rule from London initiated. The Provos kept up the pressure and ultimately in July bombed their way to the bargaining table. Provo-Cabinet talks in London led to a brief truce that collapsed into a long year of attrition and continued guerrilla war. The IRA could maintain an unpleasantly high level of violence but not escalate operations. The British Army could contain the

Provos but not eliminate them. By 1972, however, London had finally geared up and was responding to Irish matters with considerably more flexibility and initiative than before, if no more success.

In Ulster the British had followed the imperial experience. After surprise had come indignation: a tiny IRA minority, acting on behalf of a minority, had attacked the British Army performing legitimate peace-keeping duties for the benefit of all. Clearly the IRA maintained support by intimidation, since most Catholics surely wanted only decent reforms and peace. The IRA, then, was illegitimate, outlawed North and South, composed of dreadful bombers who killed innocent civilians in an effort to force the loyalist majority into an alien state against their interests, in opposition to the promises of the British Government, and perhaps even against the wishes of the Catholics, beneficiaries of all sorts of British welfare programs. Thus there was surprise, shock, outraged indignation, a public determination not to deal with a small gang of violent men who used terror to win support in the pursuit of a mistaken and flawed cause. When neither indignation nor suppression proved effective, London undertook various alternative political approaches and ultimately even talked with the Provos on the cabinet level.

Well before that, in February, 1971, when the British security forces slipped irrevocably from peace-keeping into anti-insurgency, the typical two-prong British attack on the problem had been fashioned: military and security operations in the service of political initiatives that sought to erode rebel support by creating parallel options. Even the fact that these political initiatives had been so haltingly deployed was not novel, and much political ground had already been covered before prorogation in March, 1972 by committees, investigations, and consultations, ineffectual but a beginning. The end of Stormont was seen in part as a means to open the road to a regional solution that would wean the minority from an all-Ireland solution and find some middle force in Ulster. The meeting with the Provos in London was as much to allow politics a chance to replace the bomb as a serious attempt to find an accommodation with the bombers. Still, it fit in neatly with previous British attempts to pursue a pragmatic and flexible course, however unpleasant. And in Ulster the Army, hampered by the albatross of Stormont, the difficulties of operating under the television lens and within the United Kingdom, and the restrictions on excessive rigor imposed by London, had also responded as in the past. The Army viewed the IRA through a lens little different from that found in London, sought only authority to do the anti-insurgency job necessary, and urged, if quietly, only the need for political initiatives out of London.

Events in Ulster have tended to confirm the pattern evolving out of Britain's experience with imperial insurrections. Ireland may or may not be a classic colonial case, as many Irishmen contend; but the British response can be satisfactorily compared to those earlier insurrections. The elucidation of such a pattern may by now be an academic exercise in that Britain truly seems to have run out of potential imperial rebels; but the existence of one national model implies that elsewhere other models of response may be found. Certainly the next and future rebels, if they are to be wise as well as daring, should contemplate not only the course record of their chosen opponents, but also the nature of the national character reflected in the response to revolt. The rebel forewarned may well, as did several of the rebels against the Crown, respond to strategic advantage. They, like the Provos in Belfast, emboldened and encouraged by that knowledge, may fashion a strategy to fit the traditional predilections of their enemy. And on their part, the guardians of existing order might well consider an exercise in introspective analysis. Clearly, for the complete strategist, faced with the potential or reality of insurgency, it is as well to know oneself as it is to know the enemy—they may even be interchangeable.

Bibliography

Until my *On Revolt: Strategies of National Liberation* (New York: Basic Books, 1974) appeared, there was no single work focused on the various insurgencies against the British since the Second World War. A near exception is Julian Paget *Counter-Insurgency Campaigning* (London: Faber and Faber, 1967) that examines Malaya, Kenya, and Cyprus. In general, broad analysis tends to concentrate on the political emergence of new nations (Rupert Emerson, *From Empire to Nation* [Boston: Beacon Press, 1962]) or the transformation of the British Empire (Correlli Barnett, *The Collapse of British Power* [New York: Morrow, 1972]) into the new Commonwealth (Nicholas Mansergh, *The Commonwealth Experience* [London: Weidenfeld and Nicoloson, 1969]). A fascinating popular account of the process is Colin Cross, *The Fall of Empire* (New York: Coward-McCann, 1969).

There is a considerable literature on the techniques and tactics through British eyes, subsequently published in various professional journals: R.N. Anderson, "Search Operations in Palestine," *Marine Corps Gazette,* April 1948, and Brigadier Heathcote, "Radforce-Lecture," *Royal United Service Journal,* January 1966, but seldom an overall view; rather, generalizations extended from a specific experience. Thus the most readily available sources on the British experience are studies by scholars, or by those involved, focused on a single emergency.

Palestine:

The literature on all aspects of the Palestine problem is massive (see my bibliography in *The Long War, Israel and the Arabs Since 1946,* [Englewood Cliffs: Prentice-Hall, 1968]). There is no definitive history of the Irgun, but the best study of the Mandate is still J.C. Hurewitz, *The Struggle for Palestine* (New York: Greenwood, 1968). On the rebel side, see Menachem Begin, *The Revolt* (London: W.H. Allen, 1951) and Samuel Katz, *Days of Fire, The Secret Story of the Making of Israel* (London: W.H. Allen, 1968). On the British side there is R.D. Wilson, *Cordon and Search* (Aldershot: Gale and Polden, 1949). A recent survey of the entire period is Dan Kurzman

Genesis 1948 (New York: World, 1970) which contains interesting material on the Irgun.

Malaya:

The most famous study to come out of the Malayan emergency is Sir Robert Thompson's *Defeating Communist Insurgency* (London: Chatto and Windus, 1966). Two other studies focused on the British military response are Richard Clutterbuck, *The Long, Long War: The Emergency in Malaya 1948-1960* (London: Cassell, 1967) and Edgar O'Ballance, Malaya: *The Communist Insurgent War 1948-1960* (Hamden, Conn.: Archon Books, 1966). Another interesting study is that by Riley J. Sunderland, *The Communist Defeat in Malaya* (Santa Monica: Rand D-9830-ISA, 8 March 1962).

Kenya:

There is not a satisfactory study of the Kenya emergency. The most famous is Fred Majdalany, *State of Emergency, The Full Story of Mau Mau* (London: Longmans, Green, 1962). The basis of the British interpretation can be found in F.D. Corfield, *Historical Survey of the Origins and Growth of Mau Mau* (London: Her Majesty's Stationery Office, 1960) and countered in Carl G. Rosverg and John Nottingham, *The Myth of 'Mau Mau' Nationalism in Kenya* (New York: Praeger, 1966). An interesting early work by Frank Kitson, presently the British expert on low-intensity warfare, is *Gang and Counter-gangs* (London: Barrie and Rockliff, 1960).

Cyprus:

The most interesting, perhaps even the fairest, survey of the EOKA emergency is by Charles Foley, editor of the *Cyprus Times* during the Cypriot crisis, *Island in Revolt* (London: Longmans, 1962), later extended as *Legacy of Strife, Cyprus from Rebellion to Civil War* (Harmondsworth: Penguin, 1964). On the rebel side there is Grivas in *Guerrilla Warfare and EOKA's Struggle* (London: Longmans, 1964) and *Memoirs* (London: Longmans, 1964). For a more jaundiced view of EOKA see Dudley Barker, *Grivas: Portrait of a Terrorist* (London: Cresset, 1959).

South Arabia:

A remarkable number of the British governors or high commissioners have written their memoirs—Sir Bernard Reilly, Sir Tom Hickenbotham, Sir Charles Johnston, Sir Kennedy Trevaskis (*Shades of Amber: A South Arabian Episode* [London: Hutchinson, 1968]) and Lord Trevelyan; but the most interesting if intemperate work is Colin Mitchell's *Having Been a Soldier* (London: Hamish Hamilton, 1970). Two general surveys are Julian Paget, *Aden 1964-1967* (London: Faber and Faber, 1969), and Tom Little's *South Arabia, Arena of Conflict* (London: Pall Mall, 1968).

VII

Some Conclusions about Military History

The Nature and Scope
of Military History
MAURICE MATLOFF
page 386

Military Leadership
and the Need
for Historical Awareness
HAROLD K. JOHNSON
page 412

COMMENTS

Maurice Matloff, Chief Historian of the Department of the Army Center of Military History, appropriately was the first guest speaker in the first term of the New Dimensions course, in the fall of 1971. Beginning a new course in military history, he chose largely to look ahead, to the future of military history. This approach makes it appropriate in turn to use his remarks now, at the close of our book, as a summing up and a point of departure for what still lies ahead in the development of military history. Some of the lines of research for which he called in 1971 have begun to be worked; but for the most part, Dr. Matloff's discussion of new areas of research opening up in military history and of work still to be done remains an examination of problems unsolved and challenges still to be met.

Beyond his official duties in the encouragement of military history, Dr. Matloff is at work filling in some of the missing pages he here notes as existing in the record of American military history; he will fill part of the gap in the history of strategy with a study of post-World War II American strategic thought. He has already contributed to the history of American strategy two volumes of official history, as coauthor, with Edwin M. Snell, of Strategic Planning for Coalition Warfare, 1941–1942 *(Washington: Office of the Chief of Military History, 1953), and author of* Strategic Planning for Coalition Warfare, 1943–1944 *(Washington: Office of the Chief of Military History, 1959), both volumes in* The War Department *subseries of* United States Army in World War II.

The Nature and Scope of Military History

MAURICE MATLOFF*

WE NEED to raise again the basic question, "What is military history?" The famous military historian of the British Army, Sir John Fortescue, wrestled with the problem of definition. It took him forty-six pages to deal with the ramifications. He could not come up with a single all-inclusive definition. He set forth a number of definitions. He called military history *inter alia* "the history of wars and warring," "the history of the strife of communities expressed through the conflict of organized bands of armed men," and "the history of the external police of communities and nations." And he also went on to say, "Military history is not the history of physical but of moral force, perhaps almost of the triumph of moral over purely physical force." Fortescue's reflections indicate some of the variety and range encompassed in the term. Military history is all that and more. The definition has expanded as military affairs, broadly considered, have come to occupy more and more of man's energies, either to fight wars or to deter them.

What then is military history? Each period of the past must be examined to arrive at a definition as to what it meant. It must be studied in context. In the eighteenth century, the century when the American Army was born, and before the French Revolution introduced the concept of the "nation in arms" and before war became democratized, the definition was used to

*The views of the author do not purport to reflect the position of the Department of the Army or the Department of Defense.

SOME CONCLUSIONS ABOUT MILITARY HISTORY

denote the history of conflict in arms, battles, and campaigns. Warfare, of course, in the eighteenth century was relatively simple. The area, the forces, the objectives were relatively limited. But with the spread of men in arms, with the increasing conscription that came with the French Revolution, and with the extension of the Industrial Revolution, warfare became more pervasive. It had a greater and greater impact on society, so that in the world conflicts of the twentieth century, wars were beginning to affect not only the combatant nations, but countries beyond the borders of those actually involved in the fighting. So the concept has broadened. Military history lies on the frontier between military art and science and general history. Not that military affairs should be regarded both as art and science. Military history deals with the confluence and interaction of diplomatic, political, economic, social, and intellectual factors and military trends in society. That interaction and confluence must be pitched in the broad stream of history.

In the American context military history represents many interrelated factors. Certainly it includes consideration of wars—wars of all kinds. Americans like to think of themselves as a peaceable people who abhor war. Yet our national history indicates a wide gamut and a frequent incidence of warfare. In fact, a recent writer on the subject, Professor Peter Paret of Stanford University, has suggested that the term "America the Beautiful" may really not be an apt description, that in the perspective of warfare, the nation may be regarded, at least by outsiders, as "America the bellicose."

In any event, the definition in the American context involves various types of wars. We were born in a revolution, often termed the first of the peoples' wars in modern times; we endured a very bitter civil war; we have engaged in seven international wars since independence, including three world conflicts. We have had any number of pacification experiences in the nineteenth century and beyond. The American Revolution is a good example of a limited war of the eighteenth-century variety. There are other models of limited war in our history, for example, the War of 1812 and the Spanish-American War—aside from the experiences of the more recent period. The Civil War is usually regarded as the first of the modern wars, introducing the age of total war to which belong World War I and World War II. Yet, we seem to be returning to more primitive forms of warfare. In spite of the concern of theorists with thermonuclear war and limited nuclear war, the actual fact has been the return to earlier forms. We seem to be coming full circle, and there is greater interest than ever among scholars of military affairs in such experiences as the Philippine Insurrection of 1899 to 1902 which followed the Spanish-American War. Interest is growing too in studying the Indian Wars, as examples of guerrilla and counterguerrilla action. Certainly in terms of warfare, there are any number of ingredients in our history that have to be considered.

When we talk about wars, we have to keep in mind that there has been an important shift. Wars used to be regarded as clearly definable exercises in violence. They were marked by formal ceremonies—a declaration at the beginning, a surrender, a truce, and a peace treaty at the end. In the period since World War II this formality has gone by the boards. These formalities are no longer observed. There has been a blurring of the formerly distinct periods of what we used to regard as war and peace. More and more historians are looking into periods before wars break out. And, of course, more and more of the important decisions are being taken before hostilities begin. As a result, in considering wars, military historians of this generation are considering what goes on before the outbreak of hostilities and during so-called interwar periods.

Besides war in a broad sense, another major factor that interests military historians of this generation concerns the question of armies as institutions. The same applies to navies and air forces as well. As institutions, armies reflect national cultures and have an impact on them. There are examples in history of armies that have existed as a class apart from the remainder of society. In most of our nation's history, the Army actually

SOME CONCLUSIONS ABOUT MILITARY HISTORY

existed physically apart from the rest of society. In the nineteenth century, for example, the Army was largely isolated on frontier posts. This is a factor which has to be considered any time the role of the military in a particular society is studied. As institutions, armies take form and character and the institutional outlines are revealed in a number of ways—the manner in which they select their leaders and draw upon enlisted men, the modes of supply, the methods of fighting, and the pattern of civil-military relations. A shift in one phase of the institutional structure will have marked reverberations throughout. A fundamental change in weaponry, equipment, or technology—for example, the introduction of the long-range infantry rifle, the tank, the airplane, even of gunpowder, or, in more recent times, the atomic bomb—will have marked changes upon traditional modes of fighting, upon civil-military relations, and throughout the institutional framework. In this connection the phenomenon of cultural lag, so inherent in other forms of human institutions, must also be taken into account. There is particular danger if armies lag, if military ideas do not keep pace with changes in weaponry. Military history is replete with examples of armies that attempted to apply outmoded doctrine and met with disastrous results in the field of battle, the ultimate test. The factor of cultural lag, then, should be noted in the study of military history.

As social entities armies have evolved to carry out their primary mission, to fight. But as institutions, armies must also be considered as social forces in times of peace. Our own Army, like that of other countries, has played a very important role in our country in peacetime. From the very beginning our Army participated actively in the settlement of the country, in exploration, in guarding the frontier. It made many contributions in medicine, sanitation, flood control, transportation, communication, and engineering. This role, sometimes forgotten, may become more important in the period ahead. Along with this domestic role, our Army has played a part in the upward social mobility of certain disadvantaged groups in the population. In the 1840s and 1850s, for example, European immigrants were able to climb the ladder using the Army as the base, a form of acculturation in American society. In the 1950s and 1960s the same general phenomenon was repeated with respect to blacks in the Army. The Army has really led American society in the acculturation of the Negro. Despite whatever problems they may be having in this regard, the armed forces are generally regarded as having been ahead of American society. In this connection, one of the outstanding students of this phase of military history, Professor Russell F. Weigley, regards the American Army as a blend of professional civilian elements. He sees the evolution of that Army as the history of a twofold institution—a Regular Army of professional soldiers and a civilian army composed of such elements known variously as militia, Organized Reserves, National Guards, and selectees. The blending of the two he regards as making up the unique institution known as the American Army.

We must recognize that every generation rewrites history. It looks to the past for inspiration, wisdom, and alternative courses of action. The problems in each generation determine what will be the composition of its history—with what problems the historians will be dealing. In our own country a combination of factors has tended in most of our history to play down the importance of the military factor. Of course, after the birth pangs of the American Revolution there was much concern about whether we could survive as a nation. But during most of the nineteenth century we had the benefit of what has been called "free security." We benefited by the presence of the British Navy in the Atlantic and by our geographical isolation from the turmoils of Europe. We had weak neighbors on both borders, north and south. So we could develop domestically and not worry about the military factor. To be sure, there were wars in the nineteenth century; but, with the exception of the Civil War, they were relatively minor events in the nation's history at large. This isolation, combined with the liberal tradition, promoted the feeling that the military factor was not too important

SOME CONCLUSIONS ABOUT MILITARY HISTORY

for Americans. And scholars reflected this reaction. Even in recent times the tendency persists—for example, even today on many campuses—to discount the military factor. But a great change has come over the general American outlook on its national security. George F. Kennan, the expert on Soviet-American relations, has written eloquently of the growing insecurity of the American public in this century. A country which at the outbreak of the twentieth century felt very secure and was not concerned about external threats, he pointed out, by the middle of the twentieth century had begun to feel insecure and to show concern over the foundations of its national security. For the first time since its birth national survival had again become an issue for the United States. As a result, Americans are taking a fresh look at their military past, and there has been growing interest in the military factor in our day, despite the continuance in many quarters of an anti-military feeling and a tendency to downgrade the importance of the military factor in our history.

Accompanying this shift has come a great change, particularly since World War II, in the content of military history, a transition from the "drum and trumpet" school that emphasized battles and campaigns to the "ecological" school concerned about putting warfare in its proper political, economic, and social context, with the broadening of military affairs and its larger impact upon American society. Of course, the military factor in the American experience can be applied to other societies as well in the modern world.

So far we have been talking about the substance of history, but history is many things. It is not only the study and record of the past, it is also a tool of research. It is a tool that has been used by the classical writers on strategy, for example, Clausewitz, Jomini, and Mahan; by modern strategic analysts such as Liddell Hart and J.F.C. Fuller; and even by some of the current writers on strategy, so-called scientific strategists such as Herman Kahn. Some present-day analysts, Bernard Brodie for example, use history very well. Some political scientists use it better than others. But this classical tool of research has been submerged in recent years because of the preoccupation with newer approaches such as operational research, systems analysis, and budget management. This is not to take anything away from the other tools of research. But if you look at the work of such scholars as Samuel P. Huntington, for example, one of our ablest political scientists, you find that when he came to analyze the military factors in American society, he felt he had to go into history to dig up his examples to find the trends and explore them. While not all historians would agree with his use of history, he makes rather effective and provocative use of it.

History is not only a tool of research, it is also a laboratory of experience and of social science. To the military man, this is particularly important, because in his training and in his education he needs to acquire a vicarious experience of war. He may throughout his career be confronted with only a single direct experience of armed conflict. Bismarck once said that he preferred not to learn through his own experience; he would rather learn through other people's experience. He thought only fools learned simply through their own experience. For the military, in particular, history represents a broad foundation, a gamut of experience which can be drawn upon for individual education and development. As a source of wisdom and inspiration, and a record of alternative courses of action, history offers further interest to the military.

What is the historian trying to do? The historian attempts basically to reconstruct the past. The military historian works often with the very recent past, something that his academic colleagues do not always like to do. But the military historian has found he must deal with the recent past. The historian deals with time sequences and change. He has to have pegs to hang his hat in time; that is why he uses dates. He deals with a flow of change, a flow of relationship between man and his environment, but through time. How does he do this? He uses documents, around which the craft grew up. He deals with primary and secondary sources.

SOME CONCLUSIONS ABOUT MILITARY HISTORY

A primary source is contemporaneous with an event; it could be a diary, a letter, or a memorandum written at the time, or a report written immediately after the action. In using primary sources the historian has to be sure to come up with the actual facts, with what facts were actually relevant to the questions that he is putting to the evidence. Sometimes the contemporary document will assume matters of which the later generations will not even be aware, for example, the factor of weather. If it is an action report, it may not even discuss the enemy position because it did not have the evidence. But when a scholar comes along later and receives access to the records of the other side of the hill, the opposing side, he can tell the fuller story. And what he tries to do in reconstructing the event is to portray as full and as accurate a picture as possible. In the process he will know more than what was available to the commander at the time. In constructing his story he will undoubtedly also use secondary sources, meaning accounts written after the event. Books, for example, that scholars are presently putting out about the American Revolution are secondary accounts.

In using his sources, the historian applies rules of evidence and becomes the judge of the evidence. His job is to reconstruct the events, to get as many of the relevant facts as possible, but he is not the judge or the prosecuting attorney of the actors. He is the judge of the evidence, and that is a great difference.

In recent years a new technique has grown up which military historians are using quite actively. This is the use of oral history to fill the gaps in the written record. One of the most frustrating aspects of working in high-level World War II history is the task of tracing an action from the time an Army planner in the Operations Division of the General Staff first conceived of a plan all the way up through levels of the Joint and Combined planners to the Joint and Combined Chiefs of Staff and even to the great international conferences. But when the action reached the President for decision, he might have lifted a phone and called a high official, and no record was kept. The trail disappears somewhere in the neighborhood of the White House. Sometimes oral history can fill such a gap.

NATURE AND SCOPE OF MILITARY HISTORY

A very active oral history program of debriefing key general officers is now being conducted at the Military History Research Collection, in conjunction with the Army War College, and this will fill an important gap in the record in the years to come for a post-World War II period.

We must remember that the historian has to select among his evidence. Even if he had all the facts and tried to put all those facts before you, it would be an unintelligible mess. The chances are that he will never have all the facts. Documents do not normally reveal all, and if he is using oral testimony, the historian is dealing with fallible human memories. In the process of selecting, he is evaluating and interpreting. This is where the artistic part of the craft comes in. How does he select? He has to ask questions. Carl Becker, one of the most famous American historians in recent times, once said that you can tell a good historian by the questions that he puts to the evidence. There are some practical approaches that can be used; for instance, what would be useful to the man of action and the man of thought to know about a particular action? The questions are drawn out of the individual historian's experience, reading, and training. So, basically, what the historian does is to try to bring order out of chaos. He is trying to make a pattern and to show relationships. Sometimes the resultant picture can become too orderly and artificial. The reader must beware.

The historian knows that he is dealing with unique phenomena. No two events are precisely alike. It is very important to realize this. You probably know the old adage that history does not repeat itself, but that historians repeat each other. Many myths have been perpetuated in history. For example, one that has been exploded recently is the myth that Theodore Roosevelt was mainly responsible for ordering the fleet to the Philippines at the outbreak of the Spanish-American War, that he just happened to be in the office of the Secretary of the Navy when the latter was absent one day and took the fateful action on his own. We now know that the Navy had contingency plans years before the event, and whether Theodore Roosevelt had been in the office of the Secretary or was out that day probably would not have made any difference. The plan would have been put into effect. Yet many textbooks in American history repeat the myth. Only gradually do such myths get corrected. Many other examples might be cited. The professional historian realizes that the reader studies not what actually happened, but what historians say happened. A factor is interposed between the event and the narration, and that is a scholar who has been at work.

For example, take the essay by Michael Howard on "The Use and Abuse of Military History." Michael Howard is one of England's foremost military historians, who has worked on the official British series on grand strategy of World War II and for many years was Professor of War Studies at the University of London before becoming Defense Fellow at Oxford. In this essay he called attention to three general factors that are useful to keep in mind in the study and use of military history. One is "width." This is the question of perspective, that when you read in history, particularly military history, it is important to read as broadly as you can. Get a sense of the continuities and the discontinuities. What seems to stay the same? What seems to change? Temper the principles of war, for example, with a sense of change from period to period, and compare the conditions in which they were put into effect in one period vis-à-vis another. Howard cites the example of the attempt to apply Napoleonic methods to World War I as an outmoded use of doctrine that a reading of history might have corrected. He mentions also the factor of "depth." Here he has in mind the need to get behind the seemingly orderly pattern of an event or of an action as you read it in history books. Those neat little arrows that seem to show how everything was very orderly on the field of battle are often an artificial representation. What seemed to be a calculated decision on the part of a commander turns out to have been anything but that when you look at all of the evidence. If you look at the after-action reports, the memorandums, the diaries, and the like, the decision often turns out just to have been a matter of intuition, and a matter of luck. You must look at historical actions in depth. Look at all kinds of evidence and then compare that evidence with what the historian has reconstructed. Thirdly, Howard cites the factors of "context." By this he suggests that historical events must be put in the proper environment. He says beware of analogy—and, of course, historians are suspicious of analogies. They are very suspicious of the idea that fixed principles operate in history. They would caution, as he does, not to ascribe to George Washington, for example, principles that could be gained only by reading Clausewitz or, for that matter, from a course at the War College. Don't take things out of context. Pitch developments against the background of their times.

As historians study wars, particularly recent wars, they are more and more convinced that if the student merely keeps an eye on what goes on in the battlefield, he is viewing only the top of the iceberg. Very often success or failure on the field of battle is accounted for by the will and strength of the society at home, politically, economically, socially, and so forth. Take the case of Japan, which in World War II won many battles but could not win the war. The difference between the mobilization strength of the United States and of Japan would help account for the result. If you keep an eye on the battlefield itself, you are missing a good part of the story. So the question of context and of environment is important in pitching historical events in their proper setting.

The military historian often works with current history and, of course, the agony and the ecstasy of current history is trying to find the current. When the historian works so closely to events of his own time, it is very difficult sometimes to know what is important and what is not, and to try demands all his initiative and resourcefulness. Very often he has to generate the records, and to record events and contemporary impressions of them, as they develop. Otherwise, they will be lost forever. This is a relatively new role that the historian is playing and is one of the new developments in the field of contemporary military history.

SOME CONCLUSIONS ABOUT MILITARY HISTORY

Let me turn now to another question of great moment—what is the importance of military history to the Army in the 1970s? Certainly the needs of the Army for military history are growing and will continue to grow, notwithstanding austere budgets in this area. There will be the need to catch up with recent experience, the variety of recent involvements in combat, World War II, Korea, and Vietnam. Just as officers have not had a chance to catch their breath in the post-World War II period, so the historians have not either. Of course, historical studies in the light of our involvement in Vietnam, not only the beginning and the actual conduct of the conflict but the getting out of it, will be important subjects for future officers. All that can be learned about the tactics, the strategy, the civil-military relations, and the disengagement will be of interest.

Especially, there remains the question of how wars end. We have many studies in military history related to the three C's—the causes, the conduct, and the consequences of warfare. We know a great deal about origins and causes of wars and about how they were fought. We know very little, and there has been very little good writing, about the question of how nations leave wars. This is one of the themes with which this generation will have to deal. Of course, we never had as much trouble getting out of a war as we have had in our present experience. Again, this is an indication of how present-day problems make the historian look back on the earlier experience to try to cull the record. The political scientists have been somewhat more alert than the historians and are beginning to delve into this field. For example, Professor William T.R. Fox, the Director of the Institute of War and Peace Studies at Columbia University, who makes good use of history, has recently written an excellent essay on how wars end. This is an area that will interest historians and soldiers more and more.

The military profession has undergone great changes since the end of World War II. The military man has not only had to be a fighter in combat, but he has also had to be a program manager as well. These functions carry forward the historic roles in American society of the military man not only as a fighter, but also as a promoter of general welfare and of domestic tranquility. In the period ahead, a period of austere budgets, the balanced picture of the historic roles of the officer in American society will have to be presented by the Army if it is to get a fair hearing from the American public. Of course, this brings up the whole question of the attitude of American society toward the military.

Throughout our history that attitude has been ambivalent. It would behoove military men to look back on past periods and apart from wars to see what the attitudes of society were and how the Army reacted. There is an important question, for example, that will soon be facing the military, the question of whether after the present conflict they will turn to a narrow or a broad professionalism. This problem has historic parallels in our earlier experience. The whole question of what happens to the military when it fights unpopular wars is an important theme now, with which the whole military establishment has to wrestle. The point is that the historian comes to the past through the problems of the present, and each generation has its own problems. What the historian is trying to do is get wisdom from the past. Some of the factors that are impinging on the Army today do have antecedents and precedents, and these will be the ones in particular that the military historian will be culling.

Despite the injunctions of great leaders in and out of the military on the value of history, as a nation of doers Americans have been traditionally impatient of the past. When you look at the writings of Walter Millis, who was a military scholar of considerable eminence, you will note that his first reaction to the coming of the nuclear age was that all previous military history could be tossed out the window. Anything in the area before 1950 was just part of the dead past. That was a first reaction on the part of a writer who had done considerable study of military experience. It was an overly hasty reaction on his part, good analyst that he was, and a more accurate reaction in terms of what the experience of this country and others in the post-World

SOME CONCLUSIONS ABOUT MILITARY HISTORY

War II period has been is that of Raymond Aron, one of the foremost French analysts in the field of military affairs. Aron's idea is that under the nuclear umbrella warfare is returning to more primitive and earlier forms. If you look at what has happened in the period since the end of World War II, it reflects more accurately Aron's rather than Millis' interpretation. And, of course, there has been a revival of interest in such events as the Philippine Insurrection, in which we became involved after the Spanish-American War. We need many more studies of limited and counterinsurgency warfare experiences, not only of our own country, but of other countries as well.

Another reason for the growing interest of the Army in military history is the comparative factor, which is becoming more and more important. We have been living, in the period since 1948, in an era of prolonged mobilization unlike anything in earlier American history. It is more like what went on in Europe in the period from 1870 to 1914. We need many more studies of European and Asiatic armies. We have been dealing with them since the end of World War II in a more extensive way than ever in our earlier history. We have been involved in NATO, for example. Yet there have been very few good comparative studies of conscription systems between one country and another and of the use of volunteer armies where they have been employed by other countries. There are very few good comparative studies, and more are needed. We need, for example, studies of European experience with military aid and advice. While European armies have been involved in this area, we know very little about their practices and results. As our activities broaden in our role as heir to many of Britain's responsibilities in the post-World War II period, we need knowledge about that kind of experience. We have not had it from our own earlier experiences and have to cull it from those of other armies. To the old adage about "know your enemy," a new one should now be added, "know your ally." We have to know more about the forty-odd allies with whom we have been doing business in this recent period.

Yet another reason for the growing need of military history is the broadening scope of military affairs in the post-World War II period. Military men will have to deal in broad areas of problems of national security, military strategy and military policy, areas that have interested scholars especially since the end of World War II. Yet there have been very few academic courses geared to these problems and to the men who will be at the cutting edge of our involvement in high-level national security decisions. Consider problems of deterrence, or problems involving a variety of uses of force in the whole spectrum of political and military factors. These are areas with which the graduates of the War College will be very much involved. Yet historically, if you look at the use of military history in the military establishment, it has been mainly on the lower levels. It has been adopted in ROTC programs and at West Point, where it has been used for training, tactical, and inspirational purposes. In the period ahead, military history will be needed more and more on the level of the field grade officers who will have to cull whatever wisdom, whatever source of inspiration, whatever alternative forms of action can be secured from a review of past experiences of this country and others.

Consider an additional factor. We do not have a really good historical work on strategy—on the strategic experience even of this country. American strategic experience has been quite broad, even though we have had very few original thinkers in strategy. We have done a lot of borrowing from abroad. We do not yet have a good account of the borrowing. We do not have a good account of the links between wars. We do not have a good account of the divergence between strategic theory and practice, the theory with which a country goes into war and what happens to that theory under the impact of war. We need many more works in this field, and this need will become more important for the present generation of military and civilian professionals.

Academic scholars have been interested in military affairs in the period since World War II. We have seen the emergence of "defense intellectuals." But, in the current climate of opinion

SOME CONCLUSIONS ABOUT MILITARY HISTORY

on many campuses of the country, to do classified research on military problems is becoming more and more unpopular. The military may have to develop their own breed of intellectuals in the period ahead. History presents an ideal tool, an ideal form of education for this nourishment.

Furthermore, the American public has become more and more conscious of the need to do something about the role of minorities. This has been an era of rising expectations on the part of minority groups. The Army needs more studies of the composition of armies. What has been the experience of armies of predominantly white countries with mixed ethnic composition of forces? We have very few good studies on the role of blacks in our Army. We have practically nothing on the role of Spanish Americans and Orientals and very little on the Filipinos. All of these studies are needs of this generation.

What I am suggesting is not a blind following of the past. The lessons of the past are probably clearest in the negative, what not to do rather than specific guidance to do this or that. But we ignore our past and, increasingly, other people's past at our own peril. Santayana's dictum that those who ignore the past are condemned to repeat its mistakes is nowhere more apt than in the field of military history. But you must read and interpret the past in proper width, depth, and context, as Michael Howard has suggested, and an officer cannot know what path to follow in the future unless he knows where he has been, what path his and other armies have trod in the past.

Let me turn briefly now to review trends in the field of official history and in the academic world, to see what measures are being taken to meet the needs of this generation for military history. First, let us look at the state of official military history. There is no central historical office in the federal government. Some, not all, departments have historical offices, and they largely go their own way. The Department of Defense is the largest employer of civilian historians in the Federal government, and the Army, Air Force, Navy, Joint Chiefs of Staff, Office of the Secretary of Defense, and Marine Corps have historical sections.

NATURE AND SCOPE OF MILITARY HISTORY

The Army has the largest central historical office in Washington, known as the Office of the Chief of Military History [now the Center of Military History]. That office, which has a direct supervisory link over the Military History Research Collection at Carlisle Barracks, is really a graduate center in bureaucratic dress. It had its start in its present form in World War II, when President Franklin D. Roosevelt was approached by a number of elder statesmen of the American Historical Association to ensure the recording of what was foreseen as a great national experience. Roosevelt's letter of 4 March 1942 required executive agencies to arrange for preserving records and for relating their administrative experience during the war. As a result, officers and enlisted men who were historians began to be drawn into the work for the Army. Incidentally, Roosevelt refused to allow a historian in the White House, which resulted in a great gap. Another boost to official history in the Army was given by General Dwight D. Eisenhower as Chief of Staff. In his directive of 20 November 1947, he set the pattern for what has followed ever since. He wanted historians employed by the Army to have complete access to the record. There would be no holds barred in telling the story objectively. Historians were to follow the best scholarly standards to present accurate history. The combined action of these two leaders was to give the Army its preeminent position in the field of official military history.

To judge the state of Army official history, you have to consider the impact of the *United States Army in World War II* series. Seventy volumes have been published. When that series, now in its final stages is finished, it will number seventy-eight volumes. This series, the largest cooperative historical enterprise ever undertaken in the United States, has given official history in this country a respectable position. Until World War II the Army, which had been involved in historical work ever since the Civil War, had concentrated on documentary history. The *Official Records* of the War of the Rebellion, for example, were published by the Army in the late nineteenth century. At the end of World War I, when the Historical Division was in the General Staff, officer-historians approached Newton D. Baker,

Secretary of War, with the idea of writing a narrative account of that conflict in an elaborate series of volumes. Baker turned down the proposal. He said that it would not be seemly for the Army to write narrative history, because the accounts might have to delve into political questions and might thereby borrow trouble for the Army. When World War II broke out, the Army, lacking narrative interpretative accounts of its experience in World War I, was determined not to let this happen again. To produce the narrative, interpretative volumes of the World War II experience, professional historians, most of whom had been in uniform during World War II, were employed in Washington soon after the war.

Professor Peter Paret has criticized official history on the ground that it is too conventional. However, in using narrative techniques, the official historian was applying the accepted scholarly methods of the day to official history. The *United States Army in World War II* series also has given a boost to contemporary history that has been only a recent development in American historiography. By working so close to the events and producing accounts that so far have stood the tests of criticism, official history was able to spur the study of contemporary history in the United States. It also pioneered in oral history. Mass interviews were conducted with soldiers coming off the line in the Pacific. Captured German generals in the European Theater after World War II were given recording assignments dealing with actions in which they had participated, a form of controlled memoirs. These techniques gave a strong boost to oral history, and military history has been using it ever since.

The official history program has used the best techniques of the academic world and seeks to develop further techniques to meet the needs of the material. The official history program has foremost names today, not only in military history but in other forms of academic history. Scholars like James MacGregor Burns, Forrest Pogue, Louis Morton, and Harry Coles at one time or another worked in the official military history program and matured in it. One must remember that there was no school

NATURE AND SCOPE OF MILITARY HISTORY

of military history in the United States before World War II. In a sense, that school was created in the Office of the Chief of Military History, which has had quite an impact on the field generally. The official books on World War II are basically reference works, and their footnotes were designed as guides to the sources. A considerable number of dissertations and popular books on World War II have grown out of further research based on these leads to the sources.

The change in substance of military history—i.e., that it has broadened from the "drum and trumpet" school to the "ecological" school—is reflected even in the World War II series. Books like *The Army and Industrial Manpower*, or *The Army and Economic Mobilization*, or the strategy or technical service volumes, deal very little with the battlefield. They explore the military factor in the broader context of American society. As the World War II series winds down, this broadening continues in the current monographs and studies of OCMH. This office has to respond to the needs of the staff and the needs of the schools. It could be a study of the expansion of the Army to meet a Berlin crisis, or of the Lebanon incident. The broadening is reflected in a new series which has been promulgated, the Army Historical series. The *American Military History* volume is part of that series, as is a book on logistics from the American Revolution to Korea. That series will grow in time. The Korean War series, which is sometimes overlooked, has in it a valuable study by Walter Hermes on *Truce Tent and Fighting Front*, which deals with the interaction between the battlefield and the conference table, one of the few available studies of experience in dealing with the Communist world in winding down a conflict. Another interesting study, which may be a precedent for the future, is an across-the-board study of integration of blacks within the armed forces in the period since 1945 that has been assigned to the Army historical shop. Also projected is a Vietnam War series, which will deal with Army concerns on and off the battlefield in that conflict.

In addition, there are a number of new vehicles and re-

SOME CONCLUSIONS ABOUT MILITARY HISTORY

sources in military history that are developing under Army auspices. The U.S. Army Military History Research Collection at Carlisle Barracks represents one of the very new developments. This new institution holds great promise as an aid to the study of military history in the armed services and the civilian academic world as well—it looks both ways. There are new developments such as the establishment of professorships in military history. The Chief of Staff is very interested in the establishment of such professorships, both at Carlisle and at West Point, and they will be inaugurated in July of 1972 on an annual rotating basis. They have been set up on a two-year trial period. At Carlisle a leading scholar will be based in the Collection and will also have close relations with the Army War College across the street. This development reflects a sense of awareness on the part of the Army of a need for meaningful study of its past. Up at West Point, a committee composed of officers representing that institution as well as our own office and other agencies has been studying the Army's need for military history in the period ahead. It too is calling for a broadening of the study of military history on all levels.

One further generalization may be made about what OCMH does and does not do. It does not do what best can be done on the outside. In other words, we do not want to monopolize this field by any manner or means. One of the basic missions of the Chief of Military History is to foster the study and research of military history, both in and out of the military establishment. If outside scholars can do something better, we gladly yield. We work more on the cutting edge of events. Since we are closer to the archival sources and to the operating officials, we are in a unique position in the historical framework.

Let me dwell a moment on academic trends in the field of military history. It is important to recognize that despite the present trend of anti-military and anti-establishment sentiment in the academic world, courses in military history are growing. Before World War II, there was only one course in the country in military history, at the University of Chicago. There are now at least 110 institutions offering specialized courses of one kind or another in military history, not counting ROTC courses. This is a remarkable development, and the number is increasing. The number of Ph.D dissertations in military history has also been growing. About ten percent of the dissertations in academic history in recent years have to do with military subjects. This is quite a sizable number. Dissertations on military subjects are being produced even in universities that do not teach courses in military history.

A number of institutions are developing significant programs in the field. For example, Duke University is one of the leading centers. Duke has a sizable program; it turns out a number of graduate students and doctoral people in this field. Professors Theodore Ropp, Richard Preston and Irving Holley at Duke; Harry Coles, who is connected with the Mershon Fund for national security studies at Ohio State; Louis Morton at Dartmouth, who is editing the Macmillan series on American wars; T. Harry Williams at Louisiana State; Frank Vandiver at Rice, Peter Paret at Stanford; John Shy at Michigan; Edward Coffman at Wisconsin; and Russell Weigley at Temple University represent some of the leaders in the field. Also, there is a Chair of Military History at Kansas State University, a relatively new development.

As I see it, the official historical community has really been the cutting edge and the catalyst in this field. The impact of the Office of the Chief of Military History in this field in the postwar period has been to generate a trend, to stimulate interest in the field, and to set a standard. The hope has been that the academic world would pick up this field and make its contributions to it. This seems to be happening, and we are delighted with the development. In some fields both official and academic historians can work. Others are particularly suited for academic research.

Consider the field of biography. It is somewhat difficult for official historical offices to work in the field of biography, particularly of living figures, whereas in the academic world this is much easier to do. In a Dissertation Year Fellowship Program

SOME CONCLUSIONS ABOUT MILITARY HISTORY

that we have been sponsoring with the academic world, one of the winners for the coming year will work on a biography of Major General James Franklin Bell, Chief of Staff from 1906 to 1910. Incidentally, we know very little about those early Chiefs of Staff in that period when the country began to develop as a world power. We can indirectly foster this kind of research, but it is supplementary and complementary to our program rather than an inherent part of it.

Consider the impact of military institutions. We know from work done in our office that the business world influenced the development of the General Staff in the early twentieth century. There may also have been a reverse influence. Did the development of the General Staff as a corporate staff have an influence on the business world? Did the experience of officers in the Army carry over to the organization and management of private industrial enterprises? There has been very little academic study of such cross-fertilization. We have had very few good quantitative studies of the development of the American officer corps. Work utilizing computers at the University of Michigan is promising in this area.

We need more study, and this can more easily be done in the academic world, on the impact of wars on American society, for example, on regions, cities, and towns. As a case in point, take the impact on migration within the country during World War II. What happened as the blacks moved up from the South? What was the result, for example, for domestic housing patterns? These are areas in which the academic world can work more easily, but in which it has been moving quite slowly. Consider the questions of the generic study of war and of the military and American society, of such matters as the social origins and social attitudes of American officers. These areas that are beginning to attract academic scholarship need more work. We need more knowledge of the impact of military education on civilian education. We do not have a good history of ROTC. The historic role of military education on the civilian campuses needs study.

NATURE AND SCOPE OF MILITARY HISTORY

Within the past year two provocative essays, Peter Paret's "The History of War" in *Daedalus*, and John Shy's "American Experience in War: History and Learning" in *The Journal of Interdisciplinary History*, have emphasized the need for an interdisciplinary approach to military history. Paret points out that history traditionally has been a borrower from other social sciences. He sees the fields of psychology, economics, sociology, and quantitative analysis, as well as the use of models of simulation, as important potential contributors to the study of the history of warfare, and he calls for more interest in the academic world in military history. John Shy's essay offers an interesting hypothesis about current American military behavior, which he believes can only be understood in the light of the past. He argues that a remembered military past has traditionally guided American action. The past may be coming up against new factors in the present period, whose needs it does not quite meet, and which may lead to a great crisis in the American national personality. In effect, Shy is adopting learning theories from the field of psychology for military history, and applying insight gained from the learning of an individual to the learning of a nation about its military past. These, then, are some samples of the current academic ferment in the field of military history.

Let me conclude by noting that I see no contradiction, no incompatibility, between official and unofficial history in military history. Both use basically the same standards. There is plenty of room for both. The needs and problems of this generation are manifold, and history will inevitably be rewritten in terms of those needs and problems. In this endeavor the new military history has a part to play.

Bibliography

Suggested sources for this essay include the following: Raymond Aron, *On War* (Garden City: Doubleday, 1959) and *War and Industrial Society* (New York: Oxford University Press, 1958); Carl Becker, *Everyman His Own Historian: Essays on History and Politics* (New York: F.S. Crofts, 1935); Bernard Brodie, *Seapower in the Machine Age* (Princeton: Princeton University Press, 1943) and with Fawn Brodie, *From Crossbow to H-Bomb* (Revised and enlarged edition, Bloomington: Indiana University Press, 1973); Sir John Fortescue, *Historical and Military Essays* (London: Macmillan, 1928) and *The Writing of History* (London: Macmillan, 1926); William T.R. Fox, "Causes of Peace and Conditions of War," in *Annals of the American Academy of Political and Social Science*, 392 (November 1970): 1–13; Walter Hermes, *Truce Tent and Fighting Front* (Washington: Office of the Chief of Military History, 1966); Michael Howard, "The Use and Abuse of Military History," *Journal of the Royal United Service Institution*, CVII (February 1962): 4–10; Herman Kahn, *On Thermonuclear War* (Princeton: Princeton University Press, 1960); Samuel P. Huntington, *The Common Defense: Strategic Programs in National Politics* (New York: Columbia University Press, 1961); George Kennan, *American Diplomacy, 1900–1950* (Chicago: University of Chicago Press, 1951); Maurice Matloff, General Editor, *American Military History* (Washington: Office of the Chief of Military History, 1968); Walter Millis, *Military History* [Service Center for Teachers Publication 39] (Washington: American Historical Association, 1961); Peter Paret, "The History of War," *Daedalus*, 100 (Spring 1971): 376–396; John Shy, "The American Military Experience: History and Learning," *Journal of Interdisciplinary History*, I (Winter 1971): 205–228; and Russell F. Weigley, *History of the United States Army* (New York: Macmillan, 1967). See especially, also, the volumes of *United States Army in World War II*, published since 1947 by the Department of the Army Center of Military History and its predecessor agencies.

COMMENTS

The New Dimensions course was planned for students who were themselves soldiers. It was most useful to introduce such students to military history mainly as it is seen not by other soldiers, of whom they knew many already, but as it is seen by historians, with whom their own professional lives afforded only limited acquaintance. Yet a professional soldier's reflections on military history were included in the first two years of the course and ought to be recorded in this book as well. How is military history perceived by a highly experienced, retired professional soldier, who held military positions of the highest responsibility? General Harold K. Johnson, USA, who was Chief of Staff of the Army, 1964–1968, offered his remarks first in the autumn of 1971.

Military Leadership and the Need for Historical Awareness

HAROLD K. JOHNSON

I AM AFRAID that I am here, in part, under false pretenses. I am not a historian, despite the fact that I have an enormous interest in history and particularly military history. I brought my historical "Bible" with me; and I am sure that you are all very familiar with this book, *The Lessons of History*, by Will and Ariel Durant. I want to give just one little quotation: "Most history is guessing, and the rest is prejudice." As a speaker, I have sometimes likened myself to Dr. Douglas Southall Freeman, in all humility; I'm good for about three questions an hour, as he was. But I felt that it might be well, as a start, to offer just a little bit of perspective.

When Vietnam started, one of the first things that I did was to get Army military history teams out there, because too much of the time we have been too late and we are always trying to reconstruct what happened. If one looks at the histories—the unit histories especially—of World War II, you will discover that you can't find a "wart" in them. Everything written and shown is good; but the weaknesses, the deficiencies that existed simply don't show up. This is the result of a natural, human tendency, and something I think that we must recognize. At the same time it's terribly important that we try to do something about it, so that we bring out *all* the facts instead of just half of the facts; not show only the good half and avoid the bad half. Good history must present all sides as fully and factually as space and information will permit. At this point in time I think

SOME CONCLUSIONS ABOUT MILITARY HISTORY

historical perspective is especially important, because we are in the midst of rapid transition.

I want to raise some fundamental issues with you. At Leavenworth, I was distressed with the Air Force statement of that day, "SAC has deterred war." I couldn't quite understand how SAC had deterred war when we had lost 32,000 dead in Korea. We had had a very substantial number of conflicts around the world. SAC hadn't really deterred the war; SAC had deterred a nuclear exchange, but had not eliminated war, by any definition. So I put some faculty members there to work going back through history, trying to find out if there was a common thread to all conflict; and they didn't find any. What constituted the successful outcome of wars? And those faculty members could not identify the conditions that marked a successful outcome of war. But out of those studies came my own definition of the need for an armed force, or the purpose for an armed force, that I've since used a lot of times, and that has served me well up to now.

I think that the activity just described is illustrative of the value of going back through history to draw broad lessons from it, rather than to attempt through history to identify rules by which you might pursue a course of action, or in a more generic sense, by which you might live. One just can't get rules for today from history, and I think this is the wrong approach to the use of history. You can derive broad, general principles from the study of history.

I think the right approach to the use of military history is to try to put it in the context of its time, to analyze it and say, "Well, that's the way it worked then." Now, by comparison, what factors have we got that are the same today, and what have we got that's different? Such considerations would alter the way that we might view the lessons drawn from history.

The lesson alluded to, for me, was that the purpose of an armed force is to create, or to restore, or to maintain an environment of order within which government under law can function effectively. Now, that may sound like a pretty simple purpose, but it is the basic fact and has a real application. Where do we stand today? What is going to be the nature of our national armed force?

When I say we are in the middle of a great national transition I am not sure that very many people understand this at all. An example of this great transition is that now we have the President visiting China and Russia. We haven't had the outcry that one might have expected as little as five years ago as a consequence of this announcement. At Valley Forge this week an assembly of Supreme Court Justices from eleven states and the heads of veterans' organizations and heads of service organizations has been judging material to receive recognition from Freedoms Foundation that has come in this last year. One of the men is a Supreme Court Justice from Florida, and he said, "You know, I can't get uptight about what the Russians are going to do to us, and I can't get uptight about what the Chinese are going to do to us, but I lie awake at night worrying about what's happening to us internally." Many people share his concern today, but rarely voice that concern.

Now, what does it take to defend the country? Regrettably, in all of the argument that goes on about the modern volunteer force that we are going to have, and note that I avoid the term army, but about a modern volunteer force nobody has asked this question. And the reason that they haven't asked the question is that it would be embarrassing. And more importantly, I don't think that today anybody or any group is in a position to provide an answer with a substantial degree of credibility. What does it take to defend the country?

Well, here is where leaders of our armed forces can profit from history. What did it really take to defend the country in 1812? We didn't do very well, did we? What were the causes and conduct of the Mexican War? What might have avoided the conflict of 1861? Why did we become involved with Spain in 1898? What caused us to move toward war between 1914 and 1917? Where were we militarily in 1934, in 1939, and where, when we eventually did move, in 1941? Actually, we didn't really begin to move until almost 1943. Are we sliding back to the military doldrums period of the 1930s?

In the last three years I have done a lot of roaming around and I have been associated with a lot of different people—different kinds of people and different groups of people. It has been a whole new education for me. The attitudes and the outlooks in this society of ours are diverse indeed. And what this means is you have got to have an understanding of what that diversity of outlook has been if you're going to perform effectively, so that in studying history you're not really limited to military history. One must be concerned with military history in the context of the environment, or of the time in which particular military events took place.

Where are we now? We, as military leaders, are really in a very different role than ever before. When did the jet come? The Germans were flying some late in World War II. When did we start putting big groups of people aboard this jet? Almost 1960; the very late 1950s. I got on a jet for the first time in an airport in Kansas City in 1961 to fly to London. What can one do today? Well, from Washington one can travel to any point in the world inside of twenty-four hours just by virtue of that jet. Are we really aware of the kind of change that has made in all of our lives? And the kind of influence that it is going to have on the story of this age? No, I don't think we understand the consequences. Militarily, what can we do? We can reinforce. We can run away. We can move an analysis team to the scene. We can send technical assistance. Corporations can do the same thing. But the military must have an instant response.

Next, couple the jet age with another development that was created just a little bit later, and that is the ability to see. We can look anywhere on this globe that we can get a camera. That is what has influenced us so heavily in the case of Vietnam. People could ignore the war until they had blood on their television screen along with steak on their table for dinner. But they couldn't avoid it; they couldn't ignore it any longer. Older, more mature people, who had some kind of perspective, might try, but

THE NEED FOR HISTORICAL AWARENESS

the young who are idealistic couldn't ignore this for a moment. Now these situations obviously condition us and they condition us just enormously.

Then we add a third element that has yet to be solved, and something that will not soon be solved. And that's the matter of the great mass of information being disseminated. We get information that is visible to us through the tube, but we really get this deluge of information by reason of our ability to transmit in many medias, and we haven't learned to distinguish between the important and the unimportant. In this I think that the military services have a high order of guilt, because we get more and more people so we can handle more and more information instead of getting fewer and fewer people, which would force us to get rid of the chaff and retain the essential data. When I was a small boy, we had a machine called a sweet clover purifier. Into that machine went all of the bad seeds with the clover; the clover seed is a little tiny thing yet a big complex machine was required to separate the good clover seed from the weed or bad seed. Well, we were able to purify the clover seed that was going to be planted, but we haven't been able to purify today the information that surrounds the fundamental issues with which we must deal.

What are half facts and what are whole facts? A year or so ago I was out in Chicago for a weekend meeting. One professor got up on a platform to chide another professor and said, "It is regrettable that before we started we didn't determine the number of facts that are required to establish a truth. But it's very clear that we've had very little truth from this platform today." The business of distinguishing between the good and the bad in information is substantially the clue to determining truth. And this applies as well to history as to current issues, as most history is prejudiced in some degree in the manner in which information is pulled forward to support a particular viewpoint and other information is disregarded or discarded that is not compatible with the viewpoint that is desired as the end product. In analyzing information, who does the distinguishing? How do we draw distinctions? With all of this information now available, we haven't learned how to cope with it. And it literally suffocates us. Just think of the problems of the historians of fifty years from now having to deal with all of this material.

This is the kind of environment in which we as military leaders are operating today; with a world that is infinitely more complex, with a world that is compressed, just pushed together. We haven't considered what lies outside this earth, and what influences that might ultimately have on us. It certainly grabs the imagination; somebody standing on the moon while we watch every action during the period that the information is transmitted. All of these things will influence the way that historians are going to look at our time.

What can we look for ahead? Well, let me give you a simple illustration. As a military illustration, we have always had a great deal of contention with the Congress over grade structure, officer structure in the Army. And we have had what—11.6, 11.7 percent of our strength has been permitted to be officers? It doesn't bear any relationship to the requirements for officers; it is based on jobs that people have identified that should be filled by an officer. It is just something that has been done over a long period of time, and the debate has been over one-tenth of one percent, maybe two-tenths of one percent, in any one year. Why that kind of a relationship historically? There is no reason for it; it just grew over a long period of time. And these are the sorts of things I think that you are still going to be confronted with in times ahead. While dealing with a world different from that for which we were trained, we will still have to contend with many long-standing factors of military affairs as they have evolved in this country for many years.

Vietnam probably is a very classic example; however, it is too early, perhaps, to say that there has been only a misapplication of history. Had there been a thoroughgoing analysis, in other words, a deep, historical research study of the character of Vietnamese society and history, it is unlikely that our political leaders would have projected us in there. Perhaps our presence there is partly because Vietnam has been viewed substantially as

THE NEED FOR HISTORICAL AWARENESS

a mirror image of the United States. Vietnam is a country. The United States is a country. Therefore, in some quarters it was believed that Vietnam must be like the United States. Vietnam has provinces. Provinces are something like states. Therefore, to some, the political structure in Vietnam was viewed as something like the political structure in the United States. In fact, the social and political structure of the country is entirely different; there simply is no cohesion there. There is no polarizing influence or element that could bring that country together that I was ever able to identify. South Vietnam with its diversity of population, diversity of values really, among the elements of its population, was such that one couldn't bring it together as a nation in any reasonable period of time, so that I am not sure that it was only its history that was ignored, but the actual social structure in contrast to the perceived social structure. However, a look back at what had occurred over the whole history of Vietnam would have put a different picture on the outlook that people had about it.

Certainly, no military leader can afford to be ignorant or unaware of the history of any peoples with whom his nation is either allied or in contention.